Jacob Henry Studer, Theo

The Birds of North America

Jacob Henry Studer, Theodore Jasper

The Birds of North America

ISBN/EAN: 9783743334533

Manufactured in Europe, USA, Canada, Australia, Japa

Cover: Foto ©berggeist007 / pixelio.de

Manufactured and distributed by brebook publishing software
(www.brebook.com)

Jacob Henry Studer, Theodore Jasper

The Birds of North America

Publisher's Card.

To meet a common want, and to gratify a universal taste, the undersigned takes pleasure in placing before the public a work containing beautifully colored illustrations and descriptions of over six hundred different species of birds, comprising all that are known to exist on this Continent, including a popular account of their habits and characteristics, and embracing the general outlines of the science of Ornithology with the classification or division of birds into classes, orders, groups, and families, fully describing each of these in detail.

The drawings for the work are made from life, and uniformly reduced to one-quarter the natural size, by THEODORE JASPER, A. M., M. D., an Artist and a Naturalist, who has made the study of Ornithology the business of his life.

In addition to the illustrations of over six hundred different species of birds, the work contains, in the part devoted to the science of Ornithology, drawings of—First, the skeleton of a bird; second, a bird's wing; third, the position and form of the feathers; fourth, the terminology of a bird; fifth, about forty plates illustrating different groups of birds.

The material for the work is taken from notes made by Dr. Jasper from actual personal observation in fields and forests, continued through a long series of years, and from standard works on the general science of Ornithology, or on some of its departments.

A complete account of the birds of North America, including Mexico and Central America, is to be found in this work, the classification being so arranged as to agree with the most modern and approved systems, excluding all unnecessary technicalities and irrelevant matter.

There is no more attractive study than Ornithology, the department of zoology which treats of the structure, habits, and classification of birds. The graceful forms, movements, and habits of the feathered tribes have been celebrated in all ages by poets and artists, and have furnished the instructors of mankind with lessons of wisdom. But this admiration is not confined to the poet, the artist, or the sage; it is universal. Wherever human beings are found, the forms, the plumage, the songs, the migrations, the loves and contests of birds awaken curiosity and wonder. No similar work, containing so many beautiful and faithful pictures of living birds, and so much descriptive and scientific information, is now extant, or has ever been published in this or any other country.

The work is intended to be bound into two volumes. Two frontispiece plates are furnished to each subscriber for that purpose. There are also, at the close of each volume, an index and an exposition of the technical terms used in the work.

JACOB H. STUDER.

COLUMBUS, OHIO, November 1, 1873.

Pl. 1.

Pl. III.

THE BIRDS OF NORTH AMERICA.

Popular History.

PLATE I.

The White-headed or Bald Eagle. *(Falco Leucocephalus.)*

THIS noble bird being the adopted emblem of our beloved Republic, I introduce him first to the kind reader; and he is indeed fully entitled to a particular notice, as he is the most beautiful of his tribe in North America.

The Bald Eagle has long been known to naturalists, being common to both Continents, and has occasionally been found in very high northern latitudes, as well as near the borders of the torrid zone, chiefly in the vicinity of the sea or on the shores and cliffs of lakes and large rivers. His food consists chiefly of fish, of which he seems to be very fond, but he will not refuse, when driven by hunger, to regale himself on a lamb or young pig; he will even, "in hard times," snatch from a vulture the carrion on which he is feeding.

The ardor and energy of the Bald Eagle might awaken a full share of deep interest, were they not associated with so much robbery and wanton exercise of power, for he habitually despoils the Osprey or Fish-hawk of his prey. Of the singular manner in which he does this, Alexander Wilson, in his work on North American birds, says:

"Elevated on a high dead limb of some gigantic tree, that commands a wide view of the neighboring shore and ocean, he seems calmly to contemplate the motions of the various feathered tribes that pursue their busy vocations below—the snow-white Gulls, slowly winnowing the air, the busy Tringæ (Sandpipers) coursing along the sands; trains of Ducks streaming over the surface; silent and watchful Cranes, intent and wading; clamorous Crows, and all the winged multitudes that subsist by the bounty of this vast liquid magazine of nature. High over all these hovers one whose action instantly arrests his whole attention. By his wide curvature of wing and sudden suspension in the air, he knows him to be the Fish-hawk, settling over some devoted victim of the deep. His eye kindles at the sight, and balancing himself with half-opened wings on the branch, he watches the result. Down, rapid as an arrow from heaven, descends the distant object of his attention, the roar of his wings reaching the ear, as it disappears in the deep, making the surges foam around. At this moment, the eager looks of the eagle are all ardor, and leveling his neck for flight, he sees the Fish-hawk once more emerge struggling with his prey and mounting in the air with screams of exultation. These are the signals for our hero, who, launching into the air, instantly gives chase, and soon gains on the Fish-hawk; each exerts his utmost to mount above the other, displaying in these rencounters the most elegant and sublime aerial evolutions. The unincumbered Eagle rapidly advances, and is just at the point of reaching his opponent, when, with a sudden scream, probably of despair and honest execration, the latter drops his fish; the Eagle, poising himself for a moment, as if to take a more certain aim, descends like a whirlwind, snatches it in his grasp ere it reaches the water, and bears his ill-gotten booty silently to the woods."

Dr. Franklin is rather severe on this emblem of our National Union. He says:

"For my part, I wish the Bald Eagle had not been chosen as the representative of our country. He is a bird of bad moral character; he does not get his living honestly. You may have seen him perched upon some dead tree, where, too lazy to fish for himself, he watches the labors of the Fishing-hawk, and when that diligent bird has at length taken a fish, and is bearing it to his nest for the support of his mate and young ones, the Bald Eagle pursues him, and takes it from him. With all this injustice, he is never in good case, but like those among men who live by sharping and robbing, he is generally poor, and very often lousy. Besides, he is a rank coward; the little King-bird, not bigger than a sparrow, attacks him boldly, and drives him out of the district. He is, therefore, by no means a proper emblem for the brave and honest Cincinnati of America, who have driven out all the King-birds from our country, though exactly fitted for the order of knights which the French call Chevaliers d'Industrie."

The Falls of Niagara are one of his favorite haunts, on account of the fish caught there, and the attraction presented by the numerous remains of squirrels, deer, and other animals, which perish in attempting to cross the river above the cataract.

The nest of this species is generally fixed on a very large and lofty tree, often in a swamp or morass, and difficult to ascend. It is formed of large sticks, sods, earthy rubbish, hay, corn-stalks, rushes, moss, etc., and contains, in due time, two eggs of about the size of a goose egg and of a bluish white color. The young are at first covered with a whitish or cream-colored down and have light bluish eyes. This cream color changes gradually into a bluish gray; as the development of the feathers advances, the light blue eyes turn by degrees to a dark hazel brown; when full grown, they are covered wholly with lighter or darker brown feathers, until after the third year, when the white of the head and tail gradually appears; at the end of the fourth year he is perfect and of an appearance as seen on our plate, his eyes having changed to a bright straw color.

The Bald Eagle is three feet long, and measures from tip to tip of the wing about seven feet. The conformation of the wing is admirably adapted for the support of so large a bird; it measures two feet in breadth on the greater quills and sixteen inches on the lesser; the larger primaries are about twenty inches in length and upward of one inch in circumference where they enter into the skin; the broadest secondaries are three inches in breadth across the vane; the scapulars are very large and broad, spreading from the back to the wing, to prevent the air from passing through. Another range of broad flat feathers, from three to ten inches long, extends from the lower part of the breast to the wing below for the same purpose, and between these lies a deep triangular cavity; the thighs are remarkably thick, strong, and muscular, covered with long feathers pointing backward. The legs are half covered below the tarsal joint; the soles of the feet are rough and warty. The male is generally three inches shorter than the female; the white on the head and tail is duller, and the whole appearance less formidable; the brown plumage is lighter, and the bird himself is less daring than the female, a circumstance common to all birds of prey.

PLATE II.

This plate represents a scene which I witnessed, resting near a patch of woods, between the Scioto river and the canal, about two miles and a half south of Columbus, Ohio, on one of my shooting excursions in the month of May.

A pair of Red-headed Woodpeckers had a nest in the old stump of a decayed tree; the entrance to it undoubtedly had been made by the Yellow Hammer, as the size of it indicated, it being considerably larger than the Red-heads usually make. I had previously examined this nest; there were four eggs in it at the time. At first a male Yellow Hammer tried his best to force an entrance, but was effectually repulsed by the Red-heads. The female Yellow Hammer was during this time most indolently sitting on another stump of a broken tree, seeming not to take any interest in the doings of her mate; but some time after, perhaps pressed by the necessity of laying her egg, she too took an active part against the Red-heads, and the united strength of both finally overpowered them, and they had to abandon their nest and eggs to the Yellow Hammers, who, in their turn, after having thrown out the eggs of the Red-heads, installed themselves in the nest.

The two Nuthatches which we see in the plate were led only by curiosity; they merely wanted to see what the racket was about.

The Gold-winged Woodpecker. (Picus Auratus.)

Fig. 1. The male.   Fig. 2. The female.

Though the species, generally speaking, is migratory, yet they often remain with us during the whole winter. They inhabit the continent of North America from Hudson's Bay to Georgia; they have even been found on the northwest coast of the continent. They generally arrive in Hudson's Bay in the middle of April, and leave in September. The nations there call them Ou-thee-quan-norow, from the golden color of their shafts, and the lower side of the wings. This bird has numerous provincial appellations in the States of the Union, such as "High-hole," from the situation of its nest, and "Hittock," "Yucker," "Piut," "Flicker," "Yellow Hammer," etc. Most of these names have probably originated from a fancied resemblance of its cries to the sound of the words; for the most common cry of the Gold-winged Woodpecker consists of two notes or syllables, often repeated, which, by the help of the hearer's imagination, may seem to resemble any of them.

The Gold-winged Woodpecker builds its nest about the middle of April, usually in the hollow body or branch of a tree, at considerable height above the ground, but not always, for I found the nest of one in an apple tree, less than three feet above the ground. The female lays five or six white, almost transparent eggs, very thick at one end and tapering suddenly toward the other; the young leave the nest early, climbing to the higher branches, where they are fed by the parents. Their plumage, in its color and markings, resembles that of the parent birds, with the exception that the colors are less brilliant, and the dots appear less frequently on the breasts of the young than on those of the old birds. The food varies according to seasons, and consists of worms, berries, seeds, Indian corn, etc., and this is perhaps the reason why farmers destroy this bird whenever they have a chance.

Formerly he was classed by many of the ornithologists among the Cuckoos, which was an absurdity, as he has no resemblance to them. The tongue is constructed like that of all the Woodpeckers, and he has no resemblance to the Cuckoo, except that two of his toes are placed before and two behind; he not only alights on the branches of a tree, but most frequently on the trunk, on which he will climb up or down or spirally around it, just as his fancy may be; when on the ground, he hops, his flesh is in quite good esteem.

The Red-headed Woodpecker. (Picus Erythrocephalus.)

Fig. 3. The male.   Fig. 4. The female.

This bird is more universally known than any other bird in North America. His plumage, red, white, and black, glossed with violet, added to his numbers and his peculiar fondness for hovering along the fences, is so very notorious that almost everybody is acquainted with him. His food consists chiefly of insects, of which he destroys a large quantity daily; but he is also very fond of cherries, pears, sweet apples, and other fruit; wherever there is a tree covered with ripe cherries, you may see him busy among the branches: in passing an orchard, you may easily know where to find the earliest and sweetest apples, by observing those trees on or near which the Red-head is skulking, for he is an excellent connoisseur of good fruit; when alarmed on such occasions, he seizes a capital one, by sticking his open bill into it, and bears it off to the woods. He also likes Indian corn, when in its rich, succulent, milky state, opening with great eagerness a passage through the numerous folds of the husk. The girdled, or deadened timber, so common among corn-fields, is his favorite retreat, whence he sallies out to make his depredations. He is of a very gay and frolicsome disposition; half a dozen are frequently seen diving and vociferating around the dead high limbs of some large tree, pursuing and playing with each other, amusing the passenger with their gambols. The cry of the Red-headed Woodpecker is shrill and lively, and resembles very much the cry of the tree-frog.

Farmers generally hate and destroy him whenever they have a chance; but whether this is just or not I will leave to them. I have above remarked that he also destroys thousands and thousands of destructive insects and their larvæ, and therefore I would say to the farmer, in the benevolent language of the Scriptures, not to "muzzle the mouth of the ox that treadeth out the corn;" and the same liberality should be extended to this useful bird that farms so powerful a defense against the inroads of many millions of destructive vermin.

Properly speaking, the Red-headed Woodpecker is a bird of passage. They inhabit North America from Canada to the Gulf of Mexico, and have also been found on the northwestern coast. About the middle of May they construct their nests, which they form in the body or large limbs of trees, taking in no instance but excavating the holes to the proper shape and size. The female lays six eggs of a pure white, and the young make their appearance about the 20th of June. During the first season the head and neck of the young birds are a blackish gray, the white on the wing is also spotted with black, but in the succeeding spring they receive their perfect plumage, as in our plate. The male and female differ in nothing except that the female is a trifle smaller.

The White-breasted, Black-capped Nuthatch. (Sitta Carolinensis.)

Fig. 5. The male.   Fig. 6. The female.

The White-breasted Nuthatch is common almost everywhere in our woods and may be known at a distance by his peculiar note—quank, quank—frequently repeated, as he moves up and down in spiral circles, around the body and larger branches of the tree, probing behind the thin, scaly bark, shelling off considerable pieces of it in search of spiders or other insects and their larvæ. He rests and roosts with his head downward, and appears to possess an uncommon degree of curiosity. Frequently I have amused myself, when in the woods, imitating the voice of a bird in distress, to see who would be the first to appear, and invariably the Nuthatch made his appearance first to see what was the matter. Frequently he will descend very silently within a few feet of the root of the tree where you happen to stand, stopping head downward, stretching out his neck in a horizontal position, as if to reconnoiter your appearance, and after several minutes of silent observation, wheeling round, he again ascends with fresh activity, piping his "quank, quank," as before. He is strangely attached to his native forests and seldom forsakes them; amidst the rigors of the severest winter weather his lively quank, quank is heard in the bleak and leafless woods. Sometimes the rain, freezing as it falls, incloses every twig and even the trunk of the trees in a hard transparent coat or shell of ice; on such occasions we observe his anxiety and dissatisfaction, as being with difficulty able to make his way along the smooth surface. At such times he generally abandons the woods and may be seen gleaning about the stables, around the house, mixing among the fowls, entering the barn and examining the beams and rafters and every place where he can pick up a subsistence.

The name Nuthatch is very erroneously bestowed on this family of birds. It was supposed that they could crack the hardest nuts with their bills by repeated hammerings; soft-shelled nuts, such as chestnuts, hazel-nuts, and a few more of this description, they may perhaps be able to demolish, but I never have seen them do it. Hard-shelled nuts, such as walnuts, hickory-nuts, etc., they are perfectly incapable of breaking, as their bills are not at all shaped for that kind of work. This absurd idea may have had its origin in the circumstance that we frequently observe the Nuthatch busily searching for insects in heaps of shells of broken nuts, lying on some old stump of a tree, or around it, brought there or broken by the squirrels, whilst ignorance ascribed the broken nuts to the doings of the feeble little bird.

This bird builds his nest early in April, in the hole of a tree, in a hollow rail of a fence, and sometimes in the wooden cornice under the eaves; the female lays five eggs of a dull white, spotted with brown at the greater end. The male is the most attentive husband and supplies his beloved mate, while sitting, regularly with sustenance, stopping frequently at the mouth of the hole, calling and offering her what he has brought. At other times he seems merely to stop and inquire how she is, and to cheer up the tedious moments with his soothing chatter. He seldom goes far from the spot, and when danger appears, regardless of his own safety, he flies to alarm her. When both feed on the trunk of the same tree or on adjoining ones, he is perpetually calling on her, and from the momentary pauses he makes, it is evident that he feels pleased to hear her reply.

The female differs very little from the male in color, the black being only less deep on the head and wings.

PLATE III.

The White or Whooping Crane. (Grus [Ardea] Americana.)

In former times the Cranes were classed with the Herons, to which they bear a certain alliance, but were afterward, with propriety, separated from them, and now form a very natural division in that great class. They are all at once distinguished from the Herons (Ardeæ) by the bald head and the broad, waving, and pendulous form of the greater coverts, and the shortness of the hind toe. The Crane is found in every part of the world, but the group is, notwithstanding, limited to a few species.

Our species, the Whooping Crane, is the tallest and most stately of all the feathered tribes of North America. He is the watchful inhabitant of extensive salt marshes, desolate swamps, and open morasses in the neighborhood of the sea and large rivers. He is migratory, and his migrations are regular and most extensive, reaching from the shores and inundated tracts of South America to the Arctic Circle. In these immense periodical wanderings, they rise to such a height in the air as to be seldom observed, and form at such times regular lines in about a sharp angle, frequently changing their leaders, or the one that flies foremost. They have, however, their resting stages on the route to and from their usual breeding-place, the more northern regions; and during their stay,

they wander along the muddy flats in search of worms, sailing occasionally from place to place with a slow and heavy flight a little above the surface, and have at such times a very formidable appearance. Their cry is loud and piercing, and may be heard at a distance of two miles; they have various modulations of this singular cry. When wounded, they attack the gunner or his dog with great resolution, striking with their sharp and formidable bills. They are extremely watchful, but not shy. When alone, they are constantly on the alert, and a flock of them has always regular guards. When alarmed, they never return to the same place without sending out a number to reconnoitre. As cautiously as he avoids man, he becomes as closely attached to him, when once brought into his companionship; he learns to understand every action of his master, knows his voice and shows his satisfaction when he sees him; he not only regards him as his master, but as his friend; society seems to be a necessity to him. One that I received from Dubuque, Iowa, which was caught on the Mississippi by a trapper, and has been living with me nearly four years, was at first very ferocious and could only be approached with great difficulty, but is now perfectly tame. It became in a very short time reconciled to its imprisonment, and is now very much attached to me.

The Cranes sometimes rise spirally in the air to a great height, the mingled noise of their screaming, even when almost out of sight, resembling that of a pack of hounds in full cry. On such occasions they fly around in large circles, as if reconnoitering the country to a vast extent for a fresh quarter to feed in. At other times, they assemble in great masses, forming in regular lines and standing erect, with their bills resting on the throat, whilst one will step out, open his wings and dance in the most ridiculous way before the others—the people on the Mississippi call this "preaching;" at other times several will dance regularly around each other with outspread wings. They live chiefly on vegetable food, such as Indian corn; but readily swallow mice, rats, moles, etc., with great avidity. They build their nest on the ground, about one foot in height, and lay two pale blue eggs, spotted with brown, as large as a goose egg, but more lengthened. The Cranes, as above stated, are distinguished from the other families by the baldness of their head, and the broad flag of plumage projecting over the tail, and in general by their superior size. They also differ in their internal organization, in the conformation of the windpipe, which enters the breast in a cavity fitted to receive it, and after several turns goes out again at the same place, and thence descends to the lungs. Unlike the Herons, they have not the short side of the middle claw pectinated; and the hind toe is very short, scarcely reaching the ground. The brown Crane (Grus Canadensis) is no other than the young of the Whooping Crane.

All the descriptions of former ornithologists are exactly correspondent with the above. In a flock of ten or twelve Whooping Cranes, three or four are usually of that tawny or reddish-brown tint on the back, scapulars, and wing-coverts, but are evidently yearlings of the Whooping Crane, and differ in nothing but in that and in size from the others. They are generally five or six inches shorter, and the primaries are of a brownish cast, and their legs are also a trifle darker.

PLATE IV.

The Rail. (Rallus Carolinus.)

Fig. 1, Male. Fig. 2, Female.

The Rail, or as it is called in Virginia, the Sora, and in South Carolina the Coot, belongs to a genus of birds, of which, as nearly as can be ascertained, about thirty-nine different species are known to naturalists, and these are distributed over almost every region of the habitable parts of the globe. The general character of these is everywhere the same. They run rapidly, but their flight is slow, and with the legs hanging down; they become extremely fat, and fond of concealment, and usually prefer running to flying. Most of them are migratory and abound during the summer in certain countries the inhabitants of which have very rarely an opportunity of seeing them.

The Rail usually builds his nest in a tussock of grass; the nest is composed of a little dry grass. The female lays from four to six eggs of a dirty whitish color with brown or blackish spots; the young ones run off as soon as they are hatched; they are covered with a perfectly black down, and run about among the grass like mice.

The Rails arrive at Hudson's Bay, and other northern parts, early in June, breed there, and leave again for the south early in autumn. But it is certain that some of them remain with us, as I have caught young Rails myself in the latter part of the month of June on the Connecticut river, just a little above the town of Wethersfield, in a swampy or muddy place, called there the Wethersfield Cove. I have also been informed, by persons of credit and intelligence, in several places of the Union, that they had found nests, as well as young Rails; but what is singular, none of them had ever seen [illegible] from the old ones. The Rails, as well as the Bobolinks, are very fond of the seeds of two different kinds of reeds, which grow up from the soft muddy shores of the tide water, and are alternately dry and then covered again with four to five feet of water. They rise with an erect, tapering stem to the height of six or eight feet. They grow up so close together that a boat can only with difficulty make its way through them at or near the time of high water. The seeds are produced at the top of the plant, the blossoms occupying the lower branches of the panicle and the seeds the higher.

These seeds are nearly as long as an ordinary pin and very slender, white, and sweet to the taste; also very nutritious, as appears by the effect they have on the various birds that at this time feed on them. When the seeds are in this state, the Rails take possession of them in great numbers. At this season, as you walk along the embankment of a river where these reeds grow, you can hear the Rails squeak in every direction like young puppies. If a stone be thrown among them, there is a general outcry and a reiterated "kook, kook, kook," somewhat like that of a Guinea fowl; any other sudden noise, as the discharge of a gun, produces the same effect. In the meantime none are to be seen, unless it be at or near high water; for when the tide is low, they universally secrete themselves among the intricacies of the reeds, and you may walk past or even over them, where there are hundreds, without seeing a single one. On their arrival they are generally lean and unfit for the table, but as the seeds of the reeds ripen they fatten rapidly, and from about the middle of September to the middle of October, are excellent and eagerly sought for. Their flight among these reeds is usually low, and shelter being abundant, is rarely extended to more than from ten to fifty yards. When winged and uninjured in their legs, they swim and dive with great rapidity, and are seldom seen to rise again.

I have found them several times, on such occasions, above the water, clinging with their feet to the reeds. They are very feeble and delicate in every part, except the legs, which seem to possess great strength; their bodies being remarkably thin and compressed, measuring less than an inch and a quarter through transversely, they are enabled to pass between the reeds like rats. When seen, they are almost constantly jerking up the tail. Fluttering as their flight appears in the reeds, I have seen them at other times rise to a considerable height, stretching their feet behind them and flying to such distances that I totally lost sight of them. In the State of New Jersey, where this particular kind of reed does not grow, we find no Rails. Most of them leave the Middle States before the end of October, and the Southern States early in November, though some are found lingering in the warm marshes and over the whole winter. Numbers of them have been found in the West Indies at the time of our winter season, which makes it evident that they migrate across that part of the sea between the mainland and the islands; and why should this be impossible? As the Rail can swim and dive well and fly at pleasure, he seems to me well fitted for such an undertaking.

The young Rails, the first season, resemble the females.

Some modern ornithologists have classed this bird under the genus Gallinula; but this seems to me altogether wrong, as all Rails are destitute of a frontal plate, which characterizes the Gallinulae; they otherwise have certainly a strong resemblance to them.

The Virginia Rail. (*Rallus Virginianus.*)

Fig. 3.

This elegant little bird is far less numerous in this part of the United States than the preceding, but inhabits more remote northern regions. He is frequently seen along the borders of our salt marshes, which are rarely visited by the Sora; he breeds there as well as among the meadows that border our large rivers. He is met with in the interior, as far west as the Ohio river; also in Kentucky in the groves and wet places, but only in the spring. He feeds less on vegetable and more on animal food than the common Rail. The food of this species consists chiefly of small snail shells, worms, and the larvæ of insects that it extracts from the mud with its long bill, which is wonderfully adapted to it. On this account its flesh is much inferior to the former; otherwise, its habits, its thin compressed body, its aversion to take to the wing, and the dexterity with which it runs and conceals itself among the grass, are exactly similar to those of the common Rail, from which genus, notwithstanding the difference of its bill, it ought not to be separated.

Some people call this bird the Fresh Water Mud Hen. The epithet "fresh water" is given to it because of its frequenting only those parts of the marsh where fresh water springs rise through the bogs into the salt marshes. In such places it usually constructs its nest, which is composed altogether of old dry grass and rushes. The female lays from six to ten eggs of a dirty white or cream color, sprinkled with specks of reddish and pale purple, most numerous near the greater end. They commence laying early in May, and probably raise two broods in the season. The young of this species are also covered with a pure black down, and have a white spot on their bill, and a soft and piping note. The female is about half an inch shorter than the male, the color of the breast is paler, and a little more white on the throat and chin.

These birds, like the preceding, stand and run with the tail erect, which they frequently jerk upward; they also fly exactly like them, with the legs hanging down, but only a short distance, and the moment they alight run off with great speed.

The Song Sparrow. (*Fringilla Melodia.*)

Fig. 4. Male. Fig. 5. Female.

The Song Sparrow may be found in all parts of the United States; he is the earliest, sweetest, and most lasting singer of all the Sparrows. We may call them partially migratory, for the most of them pass to the south in the month of November; but many remain with us all winter, in the low sheltered meadows and swamps. He is the first singing-bird in spring, taking precedence of the Peewee and Bluebird. His song, resembling the beginning of the Canary's song, or perhaps rather the song of the European Yellow Hammer (Emberiza Citrinella), is very short but exceedingly sweet, and frequently repeated, generally from the branches of a bush or small tree, where he sits, chanting for an hour at a time. He is very fond of frequenting the borders of rivers, meadows, swamps, and other like watery places. He is found, with a multitude of other kinds of Sparrows, in the great Cypress swamps

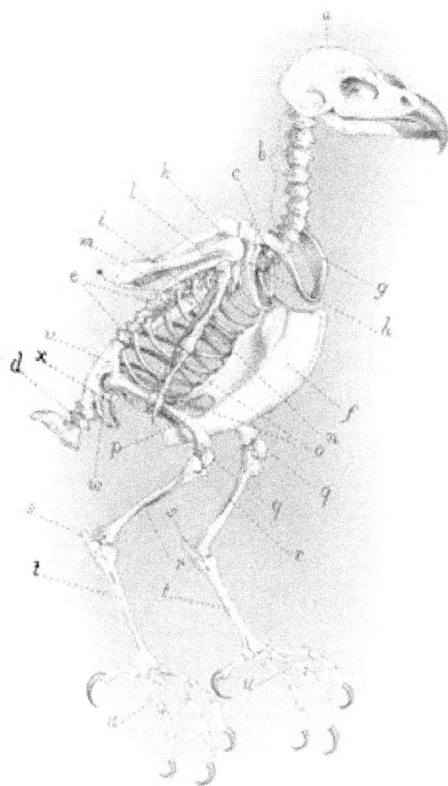

Ornithology, or the Science of Birds.

The word *ornithology* is compounded of two Greek words, *ornithos*, of a bird, and *logos*, a discourse. It is that department of Zoology which treats of the structure, habits, and classification of birds.

Birds are warm-blooded animals, and form the second class of vertebrates. They are distinguished from mammals, not only by their feathery covering, and the formation of the jaws or mandibles, which end in a horny bill, but by the production of their young from eggs, and by the formation of the anterior limbs, which, in their case, are developed into wings.

It might, at the first glance, be supposed that birds are constructed upon a plan very different from that in the case of mammals; but a little careful study of the skeletons of each will show that the two classes are built upon the same general and uniform plan.

The Bony Structure of Birds.

The following is a brief outline of the skeleton or bony structure of birds :

Explanation of Plate A. (Skeleton of a Bird.)

a. Cranium or skull. B. Cervical vertebrae. c. The anchylosed or fused vertebrae of the back. d. The caudal vertebrae. e. The ribs. f. The breastbone. g. The furcula or merry-thought. h. Clavicula or collar bone. C. Scapula or shoulder bone. A. Humerus or upper arm. i, a. The forearm (i, ulna; a, radius). s. Metacarpus or hand bones. u. Phalanges of fingers. P. Femur or thigh bone. r, t. The patella or knee pan. v, v. bones of the leg (tibia fibula). x, x. Astra calcili or heel bones. C.f Metatarsal bones. w, w Metacarpal bones. o. Ilium. n. Pubis. r. Ischium bones of the pelvis.

The Head.—The head consists of the skull and face. The former is strongly arched, and consists of several bones. These are united by seams plainly visible in the young birds; but in the old ones the bones become so compacted as to obliterate the seams. The bones forming the face are small, but peculiarly prolonged. They consist of the two upper jaw-bones and the plow-share, the quadrate and the connecting bones, together with the lower jaws. The large size of the cavities of the eyes and the thin partition which separates them are remarkable. Sometimes holes are pierced through this partition. Another peculiarity is the connection of the bones of the head with the vertebrae of the neck, which enables birds, with more ease than mammals, to move their heads in all directions.

The Vertebrae.—The backbone of a bird is divided into the vertebrae of the neck, and of the back and pelvis, and the caudal vertebrae. The number of vertebrae in the neck varies from nine to twenty-three. They are distinguished by their mobility, while the seven or eleven vertebrae of the back are immovable, and often become compacted. The same may be said of the seven or twenty vertebrae of the pelvis. The caudal vertebrae, numbering from five to nine, are more perfectly constructed than those of mammals. The last one in birds, which supports the large feathers of the tail, resembles a high three or four sided plate. The thin broad ribs, whose number always corresponds to the vertebrae of the back, are linked to these vertebrae, and with peculiar long bodies, to the breast bone. They bear, with the exception of the first and last, on their back border, hook-shaped processes, which rest on the upper border of the following lower ribs. By this arrangement the thorax is considerably strengthened.

The Breastbone may be compared to a large shield, having in the middle a ridge, varying in size and height according to the power and size of the muscles attached to it, and, consequently, according to the greater or less power of flight in the bird. In all birds of prey, for instance, this ridge is very high, and bound with

large and strong muscles; but in birds incapable of flight, it is wholly wanting. This ridge in some birds is hollow inside, to receive in its groove a part of the windpipe.

The Pelvis of birds is principally distinguished from that of mammals by its prolongation; otherwise, it has the same bones as the human pelvis.

The Merry-Thought.—The only other bone peculiar to birds is the merry-thought or wish-bone, shaped like a horseshoe, and fastened above to the collar bone, and below to the beginning of the breastbone.

The Wings.—The following are the different bones in the wings of Birds :

First. The shoulder bone.

Second. The long and strong collar bones, articulated with the breastbone, and above with the shoulder and upper arm bone, and also, internally, with the merry-thought.

Third. The humerus (upper arm bone), a long, hollow bone, filled with air.

Fourth. The ulna (elbow bone), which is usually stronger in birds than in mammals of corresponding size.

Fifth. The radius, which is weak in birds. The two latter bones form the lower or forearm.

Sixth. Two, sometimes three middle hand bones (phalanges), three fingers and a thumb. The latter has, in some birds, a strong hook-shaped nail covered with feathers. But in that case the thumb has two limbs.

Explanation of Plate B. (Structure of a Bird's Wing.)

The feathers marked from 1 to 10 are the primaries. A. The secondaries. B. The winglet, spurious or bastard wing (thumb). C. The tertiaries. A Humerus or upper arm bone. I Ulna of forearm. a. Hand bones. s. Finger bones.

Birds' Legs consist of :

First. A thigh bone (femur), and the leg bones (tibia and fibula). The fibula is very small and immovably fixed (anchylosed) to the tibia.

Second. The metatarsal or shank bones. At their lower end these bones have as many processes as there are toes. Each of these processes is furnished with a pulley for moving the corresponding toe.

Third. The toes. There are usually four of these. This number is never exceeded, while a few birds have only three toes, and the true Ostrich only two. Three of the toes are directed forward; but one, corresponding to the great toe of the human foot, is turned backward. This is the general rule. In some birds, however, the great toe is altogether wanting, or only rudimentary; in others, such as the swallow, it is bent forward. In climbing birds, such as the Parrot and Woodpecker, the outer toe and the barb or great toe are turned backward.

The Muscular and Nervous Systems.

The Muscles.—Among the muscles are those that move the wings, or the pectoral muscles. These are the most important, and are more powerful than the similar muscles in mammals. Compared with these, the muscles of the back are small. The legs have strong muscles only on the upper and lower portions, and only in such birds as have their legs feathered down to the toes.

The Nervous System very nearly resembles that of mammals. The brain exceeds the spinal nerve in bulk, but is more simple in construction. It is divided into the greater and smaller brain. It presents both hemispheres of the larger brain, but not the convolutions so characteristic of the brains of mammals. Birds, therefore, in point of sagacity, are inferior to mammals, but superior to reptiles, fishes, and insects. The bulk of the brain in proportion to the size of the body varies in the different kinds of birds. In the Eagle, it is 1-260th of the size of the body; in the Sparrow, 1-25th; in the Goldfinch, 1-27th; in the Robin, 1-60th; in the Blackbird, 1-68th; in the Canary, 1-14th; in the Duck, 1-256th; in the Goose, 1-360th.

The spinal nerve in the vertebræ of the neck is round, and of a uniform thickness. In the vertebræ of the back, it is broader and thicker, but thinner in those of the pelvis. Otherwise the nerves of birds, in their construction and distribution, greatly resemble those of mammals.

The Organs of the Senses.

All the organs of the senses in birds are well developed; single organs are simplified, but never entirely separated.

The Eye is remarkable for its large size as well as for its curious construction. Its form and size vary greatly. All far-sighted and nocturnal birds have very large eyes. Peculiar to a bird's eye is the bony ring formed of twelve to eighteen overlapping plates, which cover each other like the shingles on a roof. They differ in size, form, and strength. There is also a fan or comb, a closely folded, dark colored skin, lying at the base of the vitreous body, at the entrance of the nerve of vision, and very often extending upward to the crystalline lens. The ring and fan enable the bird, perhaps at will, to be far or near-sighted, and certainly they increase the internal mobility of the eye. Besides the two eyelids, birds possess a third, the nictitating or winking membrane. In a bird that closes its eyes, the lower eyelid is drawn up to the upper, differing in that respect from the human eye, for when the latter is closed, the upper eyelid is drawn down to the lower. The nictitating membrane is a curious appendage. When at rest, it lies in the inner corner of the eye, toward the bill; but, by the combined action of two small muscles which are attached to the back of the white of the eye (sclerotic coat), it can be so drawn out as to cover the whole front of the eyeball like a curtain. This apparatus is doubtless a very useful contrivance when the eye is exposed to a brilliant light.

Rapacious birds have the greatest range of sight. Swift-flying birds, as Swallows, which catch insects on the wing, have a very rapid flight, and an almost inconceivable quickness of sight.

The iris varies in its color according to the age, sex, and kind of bird. Brown is the prevailing color; but from that, the iris passes through all shades to red, light yellow, or silver gray; and from the last to light-gray, and even blue. Some birds have a bright green; others, a bluish black eye.

The Ears.—An external ear does not exist in birds. The great openings of the ears are located sideways on the back part of the head, and are, in most birds, surrounded by loose bunches of feathers, allowing the sound-waves to pass through without hindrance. Owls have a substitute for an external ear, consisting of a fold of the skin which the bird can open or shut at pleasure. The drum skin lies close to the entrance; the hearing passage is short and membraneous, and the drum barrel is very spacious. Instead of the three little hearing bones of the mammals, birds have one polygonous (many angled) bone, which has some resemblance to the hammer.

The Organs of Smell in birds are decidedly inferior to those in mammals. The external nose and wide nostrils are wanting. The nostrils in birds are usually located in the upper mandible, near the root of the bill, and look like little round holes or slits, either bare or covered with a skin or bristle-like feathers. The inside of the nose is divided into two cavities. In each of these are three membraneous, cartilaginous or bony muscles, covered with a mucous membrane, on which are spread the nerves of smell.

It has been asserted that the sense of smell was developed to the highest degree in the Vulturidæ (Buzzards); but experiments have clearly shown that it is the great development of the sense of sight in these birds, which draws them so quickly to the spot where a carcass is lying. But if the carcass is covered with a cloth, no vulture will come near it, be its putrid smell ever so strong. Remove the cloth from only a small portion of the putrefying mass, and the birds will come from a long distance for a feast.

Taste.—Few birds have the tongue so constructed as to serve for an organ of taste. Only the Parrot, Parroquet, and Loris, have the tongue so constituted as to justify the inference that they have the sense of taste. Their tongues are soft, thick, and covered with papillæ. A few of the Natatores (Swimmers) have a similar tongue. It is, however, far inferior to the tongue of Parrots, as these appear to select their food by the sense of taste. In most birds, the tongue is more or less separated, and is either shortened or diminished, or covered with a horny integument. In a few birds, it is long and fleshy. It is probable that the tongue is used by birds more as a feeler than as a taster, and that it may also serve to extract and take hold of food.

The Sense of Feeling, considered as mere feeling or as perception, seems to be highly developed in birds, for the outer skin is thickly set with nerves; the tongue is often endowed with feeling, and the bill covered with a soft skin.

Respiration and Digestion.

Very perfect in birds is the apparatus for the circulation of the blood, and also that for respiration.

The Heart and Lungs.—The heart of a bird has two chambers, and two ante-chambers, similar to the arrangement in the heart of a mammal, with this difference, however, that the muscles of the bird's heart are comparatively stronger than those of mammals. On both sides of the heart lay the lungs, which are rather large in proportion. Sideways of the point of the heart, are the two lobes of the liver. The lungs often adhere to the ribs, and reach farther down than in mammals. There is no division between the chest and the cavity of the abdomen, as the diaphragm is wanting. All the great bones of the limbs, and many of the body, communicate with the lungs, and are hollowed reservoirs of air. There are, also, sacs, or bladder-like receptacles, which can be filled with air, distributed about the body; some of them in the internal portions; some between the muscles and the skin, along the throat, the chest, and the sinews of the shoulders. These sacs or bladders communicate with each other, and with the lungs. The lungs can be only slightly expanded or contracted: but as a compensation for this disadvantage, the branches of the windpipe open into the lungs, and these in turn communicate with the membraneous sacs or air cells distributed throughout the entire body, so that when air is blown into the windpipe of a bird, its whole body becomes distended like a blown-up bladder.

The Windpipe consists of bony rings combined with a strong membrane, and possesses an upper and lower larynx. The former lying behind the tongue, is nearly triangular, and has no cover; its opening is surrounded by little warts, and lined on both borders by a soft, muscular skin, by which the larynx can be completely closed. The lower larynx lies at the end of the windpipe, just before the separation of the branches, and is a mere enlargement of the last ring of the windpipe. A bridge in the middle, formed by the doubling of the inner membrane of the windpipe, divides it into slits or clefts, and its borders, set in motion by the air passing out, produces the voice. The second cleft in the throat (true glottis) performs, in a great many birds, the same office as the reed in the clarinet, while the first cleft (false) acts like the ventage or small hole of the instrument, by giving utterance to the note. On each side of the lower larynx lie muscles, from one to five in number, which, by their action, may change the larynx. Their muscles are lacking in only a few birds; in others, especially in the singing birds, there are five pairs of these muscles. Besides, on both sides of the windpipe, there are some long muscles, which begin in the lower larynx, and, in some birds, run up to the ears, serving, by their action, to lengthen or shorten the windpipe. Very curious is the course of the windpipe in some birds. It does not always pass down the lower part of the neck directly into the interior of the thorax; but, in some cases, it passes before into the comb or ridge of the breastbone, and forms, with the outer pectoral muscles, a more or less deep note, turning backward and upward, and then passing down into the thorax.

The Digestive Apparatus in birds is very differently constructed from that in mammals, as the former have no teeth. Birds have salivary glands; but a mixing of the food with saliva hardly takes place, as the food is not masticated before it is swallowed. The food, in the case of a great many birds, passes first into a widened part of the throat, called a crop, where it remains and is prepared for digestion. In the case of other birds, the food passes directly into the membraneous stomach—a widening of the lower part of the throat with numerous glands, and always thinner than the third stomach or gizzard. It is never lacking in any bird, and is very large in those that do not possess a crop.

The Gizzard is variously constructed. In birds that live partially or exclusively on other animals, it usually consists of a thin, skinny sack; but in birds feeding on hard grain and seeds, it resembles a sort of grinding-mill, being composed of two semi-globular masses of thick and powerful muscles, covered on their flat inside with a strong leathery skin, and working over each other like a pair of mill-stones. This action is aided by sharp-cornered grains of sand, or little pebbles, which the instinct of the bird has prompted it to swallow for this purpose. In this way, the hardest seeds or grains are, in a very short time, ground down to a fine pulp.

The Beak.—Though the beak never performs mastication proper, yet its use varies in different birds. It is used to divide flesh, to crack nuts, to separate grain from husks, and by the aid of the tongue, to shell it like the Canary. Some birds make an approach to a sort of mastication, as the Buntings, which, by means of a knob in the middle of the palate, bruise the hard seeds before they are swallowed.

The Viscera.—In birds the larger intestine is wholly absent. A small rudimentary portion of it is found in the Ostrich. The rectum is built midway toward the termination into a sort of chamber (cloaca), into which enter the two urinary tubes, the seminal tubes, and those of the ovarium. The bile or spleen is proportionally small; the abdominal salivary glands larger; the liver is divided into several lobes, hard, grainy, and of considerable size; the gall-bladder is also large, and the kidneys long, broad, and lobed.

Organs of Generation.—Some birds have a distinct penis; all of them testes and seminal tubes. The testes are found in the cavity of the abdomen, lying in the upper part of the kidneys. In the time of mating they acquire considerable size, but soon afterward shrink to small globular bodies, and are, in some birds, hardly visible. The seminal tubes run in a zigzag way along the urinary tubes to their termination, gradually widening and forming little bladders at their termini. The grape-shaped ovaries lie on the upper end of the kidneys, and consist of a multitude of small, globular bodies, varying in number from one hundred to five hundred. The ovi-director is a long intestine-like skin, entering by one mouth into the abdominal cavity, and by another into the cloaca.

The Skin and Feathers.

The Skin of birds resembles in its formation that of mammals. Like the latter it consists of three distinct layers—the epidermis, the mucous net, and the cutis. The cutis is thin and full of folds, but becomes thicker on the feet, consisting on the soles and toes of horny scales, and undergoing a similar transformation in the bill. The cutis varies in thickness. In some birds, it is very thin; in others, thick and hard, but it is always full of vessels and nerves, and is often found with a thick layer of fat on the inner side. The mucous net lies between the cutis and epidermis, and is filled with a liquid (mucus). The epidermis consists of countless little glandular cells, of which the lower layers only are active, and filled with mucus. The layers of cells nearer the surface are more or less dried up, causing the cells to lose their round form and appear flattened, while the upper or outside layer constantly peels off. It is in the mucous net that the little feathers find their birthplace; but the larger ones pass through the cutis into pocket-shaped bags, as can be seen in the tail feathers and the larger feathers of the wings.

Explanation of Plate C. (First Ascent.)

Fig. 1. Upper or back side; the dotted places represent the comet, the blank places the convoluted parts. a. In the salivary glands, so remarkably developed in Woodpeckers. Fig. 2. Represents the same bird from the lower side or belly.

The Feathers of birds are similar to the hair, spines, or scales of mammals, but vary in the different members of the same class, as well as on the different parts of the same bird. We distinguish in the feather, the trunk and the barb or vane; and in the trunk, the quill and the shaft. The quill is the part to which the feather is fastened in the skin; it is of a transparent, cylindrical formation. Higher up it changes into a four-sided form, filled with a porous marrow, containing in its middle the so-called "soul," a row of little cells inserted into each other, and containing the nourishment of the feather. The "soul" is united or joined, above and below, to the quill.

The upper part of the shaft is arched, and covered with a smooth horny mass; the lower part is divided into halves by a longitudinal groove. On the shaft is the vane or beard, composed of a double series of layers, or thin parallel plates on each side of the shaft. Toward the shaft of the feather, these layers are broad, and of a semicircular form, for the sake of strength, and for the closer placing of them, one against the other, when in action; but toward the outer part of the vane, these layers grow slender and tapering, so as to become lighter. On their wider side, they are thin and smooth; but the upper, outer edge is parted into two hairy edges, each side having a different set of hairs, rather broad at the bottom, but slender and beaded above. In this way the beaked beards of one layer always lie next the straight beards of the next, locking into and holding each other, but offering no resistance or hindrance to the flight of the bird. Beneath these layers of feathers is the "down," similarly constructed, but more irregular and more tender. It preserves the bird from cold, to which it would otherwise have been greatly exposed.

All the feathers of a bird are divided into the outer feathers and the down. The former are subdivided into feathers of the body, coverts, wing and tail feathers. As previously observed, the wings of a bird, anatomically speaking, correspond with human arms and hands; but the human hand consists of four fingers and a thumb, while on the hand of a bird (see Plate B, n, o) there are only three fingers and a thumb, and these are all only rudimentary.

Great Consumption of Food.

The Flying Capacity of Birds.

Walking and Swimming.

Birds

North Ame

Pl. IV.

Pl. VI.

of the Southern States, which seem to be the places of their grand winter rendezvous.

The nest of the Song Sparrow is built in the ground under a tuft of grass, and is formed of fine dry grass, lined with horse-hair and other material; it lays four or five eggs of a bluish white, thickly covered with reddish-brown spots. It raises usually three broods in the season. There are young ones often found in the nest as early as the latter part of April, and as late as the tenth of August. Sometimes the nest is built in a cedar tree, six to eight feet from the ground, which seems to be very singular for a bird that usually builds on the ground; but this same habit is found in another bird—the Red-winged Starling, which sometimes builds its nest in the long grass or swamps, or in the rushes, and at other times in low trees or alder-bushes. The male and female are so nearly alike as to be scarcely distinguished from each other.

The Marsh Wren. (*Troglodytes Palustris*)

Fig. 6, Male. Fig. 7, Female.

The Marsh Wren arrives from the South about the middle of May; as soon as the reeds and a species of Nymphæa, usually called "Splatter-dock,"—which grow in luxuriance along the tide-water of our rivers—are sufficiently high to shelter it. In such places he is usually found, and seldom ventures far from the river. His food consists of insects and their larvæ, and a kind of small green grasshopper that inhabits the reeds and rushes. His notes or chirp has a crackling sound, resembling some what that produced by air-bubbles, forcing their way through mud, or boggy ground when trod upon, and can hardly be called a song. But low as he may stand as a singer, he stands high as an architect, for he excels in the art of design, and constructs a nest, which, in durability, warmth, and convenience, is far superior to the most of his musical brethren. The outside is usually formed of wet rushes, well inter-mixed with mud and fashioned into the shape of a cocoa-nut; a small round hole is left two-thirds up for his entrance, the upper edge of which projects like a pent-house, over the lower, prevent-ing the admittance of rain. Inside it is lined first with fine dry grass, then with cow's hair and sometimes feathers. This nest, when once dried by the sun, will resist any kind of weather, and is generally suspended among the reeds and tied so fast to the sur-rounding ones as to bid defiance to the wind and waves. The female usually lays six eggs of a dun color, and very small for the size of the bird. They raise usually two broods in a season.

He has a strong resemblance to the house Wren and will more to the winter Wren, but he never associates with either of them; and the last named has left before the Marsh Wren makes his appearance, which is about the beginning of September. The hired class of this little bird is large, semicircular, and very sharp; his bill slender and slightly bent; the nostrils prominent; the tongue narrow, very tapering, sharp-pointed, and horny at the extremity; and for this reason he ought to be classed—as some naturalists really have done with good cause—among the true Certhiadæ, or Creepers. His habits are also like those of the Creepers, as he is constantly climbing along the stalks of reeds and other aquatic plants in quest of insects.

PLATE V.

The Great Horned Owl. (*Bubo Virginianus.*)

Fig. 1.

This well known formidable Owl is found in almost any part of North America, from the icy regions to the Gulf of Mexico; also on the Western coast, but most abundantly in the central part of this continent.

His favorite resorts are the dark solitudes of swamps covered with a growth of gigantic timber, which he makes resound with his hideous cries, as soon as night sets in. At times he sweeps down from a tree, uttering his loud Waugh O! Waugh O! so close to you, and so unexpectedly, that you can not help be-ing startled. Besides this favorite note of his, he has other noctur-nal solos, just as melodious, especially one that resembles very strikingly the half-suppressed screams of a person being nearly suffocated; but after all, his peculiar cry is very entertaining. Another of his notes sounds like the loud jabbering and cackling of an old rooster pursued by a dog, and is kept up sometimes for half an hour. You will always take pleasure in observing him, and often, when quietly sitting under a tree, he will sweep so close by you as almost to touch you with his wings; but generally he shuns the presence of men, and seems to know that man is the worst of his enemies.

At night he is very cautious, and even in the day-time he suffers no one to approach—unlike the rest of the Owls, which allow the gunner to approach them without showing signs of being alarmed. The Great Horned Owl is rarely seen in day-time, the peculiar coloring of his feathery dress agreeing perfectly with the bark of the tree on which he sits, almost motionless. It sometimes hap-pens, however, that one of the smaller warblers discovers him, and alarms, by his cries, the whole feathered population of the forest, which now tease and keep on annoying him till he is at last com-pelled to leave his resting-place in disgust.

But it is a different thing at night; then he is bold. His flight, which, in day-time, appears rather awkward, is then silent and very swift. Sweeping low above the ground, generally, like the rest of the Owl tribe, he rises also, with ease, to great heights, and his movements are so quick that he catches regularly any bird he has scared up from sleep. Any bird—the smallest warbler as well as a chicken or a duck—will serve him for a meal; and this may ac-count for the circumstance that all birds, without an exception, hate him. He lives also on squirrels, rats, and mice, of which he destroys great numbers.

He pairs usually in February. At this time the male, after hav-ing performed the most ridiculous evolutions in the air, alights near his chosen female, whom he delights with his bowlings, the snap-ping of his bill, and his extremely ludicrous movements. This style of love-making he practices in day-time as well as at night.

His nest, which is proportionally very large, is usually built on a thick horizontal branch of a big tree, close to the trunk. It has been found in the crevice of a rock. It is composed of crooked sticks and coarse grasses, fibers, and feathers, inside. The eggs, which number from three to five, and even six, are almost globular, rough, and of a dirty white color. The male assists the female in sitting on the eggs. The young are covered at first with a thick white down, and remain in the nest until fully fledged. Even then they follow their parents for a long time and are fed by them, ut-tering a mournful, melancholy cry, perhaps to stimulate them to pity. They are much lighter colored than the old ones, and ac-quire their full plumage in the following spring.

Although the Great Horned Owl, as above stated, prefers retire-ment, he sometimes takes up his abode in the vicinity of a detached farm, and causes great havoc among the poultry, especially the young poultry, of the farmer, by occasionally grasping a chicken or Guinea fowl with his talons, and carrying it off to the woods. When wounded, he exhibits the most revengeful tenacity of spirit, disdaining to scramble away like other Owls, but courageously facing his enemy, producing his powerful talons and snapping his bill. At such times his large eyes seem to double their usual size, and he shuts and opens them alternately in quick succession as long as his enemies remain in his presence. The rising of his feathers on such an occasion gives him a very formidable appear-ance, and makes him look nearly twice as large as usual.

In former times, this Owl, as well as Owls in general, was re-garded with a great deal of superstition, and we often find the Owl introduced in gloomy midnight stories and fearful scenes of nature, to heighten the horror of the picture; but knowledge of the general laws and productions of Nature has done away with this superstitious idea, as well as with so many others. With all his gloomy habits and ungracious tones, there is nothing mysterious about this bird, which is simply a bird of prey, feeding at night and resting during the day. The harshness of his voice is occasioned by the width and capacity of his throat. The voices of all carnivorous birds and quadrupeds, are likewise observed to be harsh and hideous.

The Great Horned Owls are not migratory; they remain with us during the whole year. The female is, like all birds of prey, considerably larger than the male, but the white on the throat is not so pure, and she has less of the bright ferruginous or tawny color below.

The Rose-Breasted Grossbeak. (Coccoborus Ludovicianus.)

Fig. 2, the Male.   Fig. 3, the Female.

This elegant species of the Parrot Finches (Phytil) is found most abundant in the New England States, especially Massachusetts, but with the exception of the extreme western parts and coast of Georgia and the Carolinas, they are met with, at certain seasons, in almost every part of the United States. His wanderings extend as far up as New Brunswick, Nova Scotia, and Newfoundland, where he has been observed to breed. He leaves early in the fall to take up his abode in warmer regions and as soon as spring sets in, commences his wanderings eastward again. He is seen in Kentucky as early as the 16th of March, on his eastern travel. His flight is steady, and at a considerable height. At times he will lower himself and take a rest in the top branches of a high tree. Before taking a new start he will utter a few very clear and sweet notes. You may hear the same, at times, during his flight, but not when he is resting. At about sundown he chooses one of the highest trees to sit upon, in a stiff and upright position, and after a few minutes repose retreats into a thicket to spend the night.

His food consists of grass and other seeds, buds of trees, tender blossoms, and berries, especially those of the Sour Gum, on which he eagerly feeds; he also subsists partly on insects, which he often catches on the wing, as most of the Finches do.

In the third year he arrives at his full plumage. The younger birds have the plumage of the back variegated with light brown, white, and black, a line of which extends over the eye. The rose-color reaches to the back of the bill, where it is speckled with black and white. Our plate shows the full-plumaged female, who, therefore, differs considerably from the male.

The Rose-Breasted Grossbeak is, in common opinion, one of the sweetest singers of this continent. His song is rich and melodious, and he sings at night as well as in day-time. His notes are clear, full, and very loud, suddenly changing, at times, to a plaintive and melancholy, but exceedingly sweet, cadence. One loves to observe him on such occasions, and can not help thinking that he must himself be fully aware of his good singing talent, from his gestures and the positions he takes while pouring forth the sweet notes from the depth of his breast. In captivity he sings frequently and just as well, though not so loud.

His nest is found from the latter part of May to the beginning of July. It is fixed on the upper forks of bushes, on apple trees, or even higher trees, mostly in the neighborhood of water. It is composed of thin branches, intermixed with dry leaves and the bark of the wild grape, lined inside with dry roots and horse-hair. The female lays four eggs of a bluish white color, sprinkled with oblong specks of a brownish purple, especially at the larger end. They are hatched alternately by both male and female. The young are fed with insects exclusively, as long as they are little; then as they grow, with seeds also, which were previously soaked in the crops of the parents.

The American Red Start.   (Setophaga Ruticilla.)

Fig. 4.

This little bird has been classed by several of our best ornithologists among the Sylvicolinæ (Warblers). We will not, therefore, venture to remove him, though we would rather have him placed among the Muscicapidæ (Fly-catchers), as there is hardly any other in the whole tribe that has the characteristic marks of the genus Muscicapa more distinct than he. The formation of his bill, the forward-pointing bristles, and especially his manners, stamp him a Fly-catcher. He is in almost perpetual motion, and will pursue a retreating party of flies from the top of the tallest tree to the ground in an almost perpendicular but zigzag line, while the clicking of his bill is distinctly heard. He certainly secures a dozen or more of them in one descent, lasting not over three or four seconds, then alights on an adjoining branch, traverses it lengthwise for a few moments, and suddenly shoots off in a quite unexpected direction after fresh game, which he can discover at a great distance.

His notes or twitter hardly deserve the name of song. They resemble somewhat the words. Weese! Weese! Weese! often repeated as he skips along the branches; at other times this twitter varies to several other chants, which may easily be recognized in the woods, but are almost impossible to be expressed by words. In the interior of the forest, on the borders of swamps and meadows, in deep glens covered with wood, wherever flying insects abound, this little bird is sure to be found. He makes his appearance in Ohio in the latter part of April, and leaves again for the South at the beginning of September. Generally speaking, he is met with all over the United States, and winters chiefly in the West Indian islands.

The name Red Start is evidently derived from the Dutch " Roth Start" (Red Tail), and was given to him by the first settlers, from his supposed resemblance to the European bird of this name, the Motacilla Phœnicurus; but he is decidedly of a different genus, and differs not only in size, but in manners and the colors of the plumage.

The Red Start builds his nest frequently in low bushes, in the fork of a small sapling, or on the drooping branches of the elm, a few feet above the ground. The exterior consists of flax, or other fibrous material, wound together and moistened with his saliva, interspersed here and there with pieces of lichen; inside it is lined with very fine soft substances. The female lays five white eggs, sprinkled with gray and little blackish specks. The male is extremely anxious about them, and, on a person's approach will flirt within a few feet about the nest, seemingly in great distress. The female differs from the male, in having no black on the head and back. Her head is of a cinereous color, inclining to olive. The white below is not so pure. The lateral feathers of the tail and breast are of a greenish yellow; those of the middle tail, of a dark brown. That beautiful aurora color on the male is, so her, very dull. The young males of the first season look almost exactly like the females, and it is not until the third season that they receive their complete colors. Males of the second season are often heard in the woods crying the same notes as the full-plumaged males, which has given occasion to some people to assert that the females of this bird sing as well as the males.

The Black-Throated Blue Warbler.   (Dendroica Canadensis.)

Fig. 5.

This bird is one of those transient visitors that, at about the end of April or the first week of May, pass through Ohio, on their route to the north to breed. He reminds one, in his manners of the Fly-catcher, but the formation of his bill as well as his general appearance, places him unmistakably among the Warblers.

CREEPER—WARBLER—HAWK. 7

But little can be remarked here concerning this bird, as it is only to be met with now and then in spring, and during a sojourn of nearly eight years in Ohio, the writer has seen it only twice in the fall; but as the woods are then still thick with leaves, and the bird perfectly silent, it is more difficult to get sight of him, and he probably makes a shorter stay than in spring. Although no pains were spared to find his nest, here as well as in more northern districts, still the search has not been successful. During summer not one single individual of this species has been observed.

Our plate shows the male. The female has a kind of a dusky ash on the breast, and some specimens which had been shot were nearly white.

The Black and White Creeper. (*Mniotilta Varia.*)

Fig. 6.

This is also one of the little birds which ought to be respected by farmers and husbandmen generally, on account of his extreme usefulness. He cleans their fruit and forest trees of myriads of destructive insects, particularly ants, although he does not serenade them with his songs. He seldom perches on the small twigs, but circumambulates the trunk and larger branches, in quest of ants and other insects, with admirable dexterity. He is evidently nearer related to the Creepers than to the Warblers, for his hind claw is the largest, and his manners, as well as his tongue, which is long, fine-pointed, and horny at the extremity, characterize him strongly as a true Creeper. He arrives in Missouri, toward the latter part of April, and begins soon afterward to build his nest. One which we had the good luck to discover was fixed in the crack of the trunk of a large tree, and was composed of some fibers and dry leaves, lined with hair and a soft cotton-like down. It contained five young ones recently hatched. This was on the 28th of April. At about the beginning of October, the whole tribe leave again for warmer climates, probably the West Indies, though we have been informed that at least several of them have been perceived in the Gulf States during the whole winter.

The male and female are nearly alike in colors.

The Yellow-Throated Warbler. (*Dendroica Superciliosa.*)

Fig. 5.

The habits and manners of this splendid little bird are not consistent with the shape and construction of his bill, his ways being those of the Creepers or the Titmouse, while the peculiarities of his bill rank him with the Warblers. His notes, which are loud and spirited, resemble strongly those of the Indigo Blue Bird (Cyanospiza Cyanea). He utters them every three or four minutes, while creeping around the branches or among the twigs in the manner of the Titmouse. On flying to another tree, he frequently alights on the trunk and creeps nimbly up and down or spirally around it, in search of food, like a Creeper. He leaves the North for a short time only in winter, and can not, therefore, migrate very far South. They have been seen in the North as late as the middle of November, and as early again in the spring as the 10th of March. In the State of Connecticut, on the banks of the Connecticut river, great numbers of them have been observed as late in the fall as the 10th of October. They are rarely met with there in the spring, but why, we are unable to state. They seem to be rather partial to running waters, in the vicinity of which they are invariably found; sometimes on trees, sometimes hanging on fences, head downward, like the Titmouse, or searching among the dry leaves on the ground.

The bird on our plate is the perfect male. As to the female, her wings are of a dingy brown, and her colors in general, particularly the yellow on the breast, much duller. The young birds of the first season are without the yellow.

PLATE VI.

The Wandering Falcon, or Great-Footed Hawk. (*Falco Peregrinus.*)

The Wandering Falcon, Mountain Falcon, Rock Falcon, Duck Hawk, or Great-footed Hawk, justly deserves his names. He roams almost all over the world. His home extends from the northeast of Asia to western Europe, and the question is yet to be solved whether our American bird is a different species or not. It is evident he is not; for the size, as well as the general characteristic traits of both the American and the one described by European writers, agree almost to minuteness. Some of the European ornithologists differ somewhat in the description of his coloring; but these discrepancies seem to have been occasioned by specimens of different ages, more than by any other cause. He is also found in the interior of Africa, and, according to Jerdon, in India. This excellent observer says: "The Wandering Falcon is found throughout India, from the Himalaya to Cape Comorin, but only during the cold season: especially plentiful near the sea-coasts, or on the shores of large rivers. He does not breed there, as far as I can ascertain; nor is only a winter visitor, who appears in October and leaves again in April." In America he extends his wanderings far to the South; whether they reach to South America has not been ascertained, but it is certain that he flies across the Gulf of Mexico. To his immense faculty of flying, a distance of a few hundred miles is mere fun. He inhabits large forests, especially those interspersed with high steep rocks, but is occasionally found close to habitations, and even large cities. The one that served for our drawing was, for instance, shot in the neighborhood of Columbus, Ohio, on the Scioto river, in the month of September. He is a powerful, daring, and extremely agile bird, and experience shows that he knows, too, how to make use of his natural gifts. His flight is extremely swift, mostly close to the ground, in spring only rising to heights immeasurable and almost out of sight. He seldom is sailing but rapidly flapping his long wings. Before rising, he flies a short distance low above the ground and with expanded tail. He is very shy and cautious, choosing the densest place farthest to pass the night, and if such be too far to be reached, prefers setting on a piece of rock in an open field. His voice is strong and penetrating, sounding somewhat like Kajak! Kajak!

The Wandering Falcon attacks birds only, from a Wild Goose down to a Meadow Lark. Among Pigeons, Quails, and Grouse he makes the greatest havoc, but is especially fond of Ducks, which he pursues with untiring tenacity. Water-fowls, when approached by a gunner, usually take to the wing; not so if our Falcon is visible. Then they make off speed to the water and dive, and those only which are on land or in shallow water fly off, till they reach deep water, then suddenly drop and dive also; but this caution on their part is of no avail, for the Hawk will hover above the water till they are exhausted, then strike down upon them and pick them up.

All birds seem to know him, for not one attacks him, not even the otherwise courageous Crows. All are anxious to save themselves as soon as he is in sight. He usually strangles his prey in the air, before it can even reach the ground. Larger birds, such as the Wild Goose, which he has seized, are tormented by him in the air until they drop down with him, and then are killed. By throwing himself with full force upon his victim, the latter is stunned by the concussion, and drops. This is probably the reason he never attacks a bird that is sitting on the ground, as he would run the risk of killing himself by the concussion. Small birds he carries away to a convenient place; larger ones he eats on the spot where they dropped, plucking off some of their feathers before he begins. Small birds he devours, together with the intestines, which he rejects in the bigger ones. In his attacks he very seldom fails, and they seem to be but play to him.

His nest is chiefly built in cracks of steep rocks, difficult, if not impossible, or indeed sometimes also, in high trees. It is carefully constructed of thicker or thinner branches and sticks. The eggs, three or four in number, are laid at the beginning of June, and are of a reddish-yellow color, sprinkled with brown, more thickly so at the larger end, and the female hatches alone. The young ones are fed at first with half-digested food from the crops of the parents, afterward with different kinds of birds. When they are able to fly they are instructed by the parents in the act of hunting.

It is a well-known fact that all true Falcons, when attacked, drop their booty and leave it to the attacking party, and the beggars among the birds of prey, being aware of this, profit by it. There they sit, those stupid, lazy fellows, watching the Hawk till he has struck down a bird, when suddenly they assault him. Our hero, otherwise afraid of no bird, drops his prey at their approach, and with an indignant Kajak! Kajak! up and off he goes.

The bird of which the Hawk has taken hold in our Plate is—

The Pin-tail Duck. (*Anas—Dafila Acuta.*)

The Pin-tail is a common and well-known Duck, much esteemed for its excellent flesh, which is generally in good order. It is a shy and cautious bird, feeding in mud flats and shallow fresh-water marshes, but is rarely seen on the sea-coast. It has a kind of clattering note, is very noisy and vigilant, and usually gives the alarm at the approach of the gunner.

Some of the Duck tribe, when alarmed, disperse in all directions, but the Pin-tails cluster confusedly, giving the expert gunner a capital chance to rack them with advantage. They do not dive except when winged.

They inhabit the whole northern part of this continent, as well as the corresponding latitudes of Asia and Europe. Great flocks of them are sometimes observed on the rivers near the coasts of England and France.

Our plate shows the male. The female has the crown of a dark brown color; the neck of a dull brownish white, thickly speckled with dark brown; breast and belly of a pale brownish white, interspersed with white; back and root of the neck above black, each feather elegantly waved with broad lines of brownish white. These wavings become rufous on the scapulars, vent white, spotted with dark brown; tail dark brown, spotted with white, the two middle feathers only half an inch longer and more slender than the rest.

The other two birds on the Plate are the male and the female

Blue-winged Teal. (*Anas—Querquedula Discors.*)

The Blue-winged Teals are the first that return to the Central States from their breeding-places in the North. They arrive as early as the middle of September, and usually cluster close to the borders of the water, generally crowded together, so that gunners often kill a great number at one shot. Their flight is very rapid, when they alight they drop suddenly among the reeds or on the mud in the manner of the Snipe or Woodcock. They live chiefly on a muddy food and are esteemed kind of the reeds of marshes and lakes. Feeding on such they become extremely fat in a short time. Their flesh is excellent for the table. The frost compels them to leave in the South, for they are delicate birds and very susceptible to cold. They abound in the southern lowlands of the Southern States, where they are caught in vast numbers in hollow trees commonly called "figure two" and placed here and there on the spot rising out of the water, and strewn with rice. In April they pass through the Central States again, northward bound, making only a short stay.

PLATE VII.

The Green Heron. (*Ardea—Butorides Virescens.*)

Fig. 1.

Public opinion shows but little liberality toward this bird, having stigmatized him with a vulgar and indelicate nickname, and treating him as perfectly worthless and with contempt. This is injustice; he keeps himself as clean as any other of the whole Heron tribe, lives in exactly the same way as they do, and at the same places with them, but he is most numerous where cultivation is least known or cared for.

He makes his first appearance in the Central States early in April, as soon as the marshes and swamps are completely thawed. There, among the ditches and islands, the Sags and quagmires, he moves with great cunning and dexterity. Frogs and small fishes are his principal game, but on account of their caution and facility of escape their capturing requires all his address and quickness. With his head drawn in, he stands on the lookout, silent and motionless, like a statue, ever ready for an attack. The moment a frog or minnow comes within his range, with one strike, quick and sure as that of a rattlesnake, it is seized and swallowed in a wink. He also feeds on the larvæ of several insects, especially those of the dragonfly, which lurk in the mud.

When alarmed, he rises with a hollow guttural scream, but does not fly far, and usually alights on a fence or an old stump and looks out with extended neck, but now and then with his head drawn in so that it seems to rest on his breast. When standing and gazing on you this way, he is often jetting his tail. Sometimes he flies high, with doubled neck and his legs extended behind, flapping his wings bravely, and traveling with great expedition. He is perhaps the most numerous and the least shy of all our Herons, and is found in the interior as well as in the salt marshes.

At the latter part of April he begins to build, sometimes in single pairs in swampy woods, often in company with others, not unfrequently with the Night Heron. The nest, which is fixed on the limb of a tree, consists wholly of small sticks lined with finer twigs loosely put together, and is of considerable size. The female lays three or four eggs, of an oblong form and a pale blue color. The young do not leave the nest until perfectly able to fly.

The Cat Bird. (*Mimus Carolinensis.*)

Fig. 2.

This is a very common and very numerous species in this part of the Continent, well known to everybody. In spring or summer, when approaching thickets of brambles, the first salutation you receive is from the Cat Bird. One unacquainted with his notes would conclude that some vagrant kitten had got bewildered among the briers and was in want of assistance, so exactly alike is the call of this bird to the cry of that animal. Of all our summer visitors he is the least apprehensive of man. Very often he builds his nest in the bushes close to your door, and seldom allows you to pass without paying you his respects in his usual way. By this familiarity he is entitled at least to a share of hospitality, but is often treated with cruelty instead. It is true he steals some of the best and earliest of the farmers' strawberries and cherries, but he lives mostly on insects, of which he destroys incredible numbers. Besides, he is one of our most interesting singers. He usually sings early in the morning before sunrise, hovering from bush to bush, hardly distinguishable in the dark. His notes are, however, more remarkable for their singularity than for melody. He chiefly imitates the song of other birds, frequently with perfect success. Sometimes he seems to be at a loss where to begin, and pours out all the odd and quaint passages he has been able to collect. In un-

oiled, and their bodies are thus protected from the wet. A bird swimming on the surface of the water maintains its position without difficulty. Each stroke with its legs produces a forward movement. Generally it uses the feet only for swimming; it folds them together, and pushes them forward, spreading out the toes and pressing them back. In easy swimming, it uses first one leg and then the other; in quick swimming, it uses both legs at the same time. In steering, it lays one leg backward, with the toes spread, and uses the other leg as a rudder.

Diving is often combined with swimming. Some birds swim faster under than upon the surface of the water, and are said often to race with the fishes. Some are only able to dive by darting down from a height. The birds that dive from the surface, with a leap or spring, more or less apparent, are called swim or spring divers, and those that dart from a height. The swimming divers are skilled operators; the pushing divers are bunglers. The former easily dive deep into the water, and remain under the surface as long as they can hold their breath; the latter having forced themselves under the water by a powerful exertion, necessarily soon rise again to the surface. The former search for food under the surface; the latter only seize such prey as they may have seen from the height. Short wings are suited to swim diving, and long ones to push diving. Only one group of birds—the storm divers—combine, in a certain sense, both facilities. The swim divers use both legs and tail; the push divers principally use the wings, and some other divers use legs, tail, and wings.

No general rule can be laid down, either as to the rapidity with which birds dive, as to the depths to which they descend, or as to the length of time they remain under water. Eider Ducks are said to dive to the depth of sixty fathoms, and to remain under the surface six minutes. The majority of diving birds do not, however, descend to such a depth, and do not remain under water longer than two or three minutes. Other birds, not belonging to the class of swimmers, not only dive, but run along on the ground under the water.

Climbing and Singing Birds.

CLIMBERS.—Many birds possess a great capacity for climbing. For this purpose they chiefly use their feet, sometimes their bills and tails, and occasionally their wings. Parrots are poor climbers. They lay hold of an upper branch with the bill and pull the body up. But Woodpeckers climb more artistically, using only their feet and tail. Some birds flutter upward; in every rising movement they use their wings, lifting them up and then drawing them in. The Alps Wall-creeper (tichodroma muraria) moves in this way; while the Woodpecker climbs with a sort of hop, without moving its wings. Climbers generally move upward or sideways on the upper side of a limb or branch; but some move downward or forward on the sides of the branch.

SINGING BIRDS.—Most birds have loud, full, and clear voices, though with many the voice is a shrill, unpleasant squeal. Mute birds are unknown. For their different sensations and impressions, birds have peculiar sounds or notes, forming a kind of language which they understand, and which may, to some extent, be learned by a careful observer. By these sounds and notes, they call each other; they express their love or hate, their joy or grief; they challenge each other to fight; they warn against approaching enemies, and, in short, make an almost endless variety of communications. Not only do individuals of the same species understand each other, but those better endowed communicate with the less-gifted ones. To the warning cry of one of the larger marsh-birds, listen all the smaller rabble in that locality, and by the alarm-cry of the Robin, all the wood-birds in the forest are warned. The tell-tale Tattler sounds an alarm and warns the Ducks of the approaching gunner. The more cautious seem often to act as guardians of the entire community of birds. In the season of courtship

or love-making, birds converse, talking, singing, and caressing. Parent birds speak and sing tenderly and affectionately to their offspring. Some appear to practice in concert, talking to and answering each other. Others seem merely disposed to give expression to their feelings, not caring whether they are understood or not. To these belong the singing-birds, the pets of the creation. In communicating with each other, both sexes appear to be on equal footing; but the males only are endowed with the privilege of song. Sometimes, but seldom, the females learn to sing a few notes.

With all the singers proper, the muscles of the lower larynx are, as a general rule, similarly developed, yet their power of song is very different. Each separate kind has its peculiar keys and a certain volume of sound; each combines these keys or notes in trills, which are easily distinguished by their greater or less fullness, volume, and power. Some songs are composed of only a few notes, while others contain octaves. Besides, each singing-bird has numerous variations in its song. The same bird has one song on the mountain and a different one on the plain, though the difference can only be distinguished by close and continued observation. A good singer in a certain district may educate inferior ones to sing well, and a bad singer may spoil a good one. Some birds, not satisfied with their own natural song, mix with it the songs of others, and even striking sounds and noises. These we call Mocking-birds. Singing-birds are found in all the zones of the earth, but are most numerous in the temperate zones.

Range of the Senses.

SIGHT.—The structure of the eye enables a bird to command a wide range of vision. Within this range it observes any object with incredible quickness. Birds of prey perceive resting or flying insects at an astonishingly great distance. Their eyes more constantly, as the proper focus must be found for each distance. This may be proved by a simple experiment. If the hand is moved toward the eye of a bird of prey, the pupil expands or contracts in proportion as the hand is brought nearer to or moved farther from the eye. This shows why these birds see the smallest objects, when sailing thousands of feet above the earth.

HEARING.—The singing of birds proves their acute hearing. Shy birds become aware of approaching danger chiefly by the sense of hearing; domesticated birds attend to the feeblest call. Birds that have a sort of external ear, doubtless use their hearing as well as sight; but, as a general rule, it is probable that the hearing of birds is by no means as acute as that, for instance, of the bat, the cat, or any ruminating mammal.

SMELLING.—Though the sense of smell is but feebly developed in birds, yet a certain amount can not be reasonably denied them.

TASTING.—The sense of taste in birds is dull compared with that of mammals. Birds prefer certain kinds of food, and reject other kinds; but this can not be ascribed altogether or chiefly to the discriminating nature of their sense of taste, since, with few exceptions, they swallow their food without any sort of mastication, and many birds seem to use the tongue more for probing for, than for tasting their food.

FEELING.—Birds, in most cases, evidently use the tongue as an organ of feeling. Woodpeckers, Humming-birds, and many others probe with the tongue the hiding-places of their prey, and by the same organ separate the digestible from the indigestible portions of their food. The general development of the sense of feeling in birds increases the keenness of their perception, and guards them against sudden changes in the weather and other external influences.

Distribution of Birds.

As far as the surface of the earth has been explored, birds have been found in high northern latitudes, as well as between the tropics; on the plains, in the valleys, on the seas, and on the highest mountains; in fertile regions, and in deserts; in primeval forests, and on the barren rocks which are elevated above the surface of the surrounding waste of water. Each separate zone has its peculiar feathered inhabitants. Birds in general follow the laws of animal distribution. They are found in great numbers in the frigid zones, but are represented only by a few kinds. Toward the torrid zone, the number and varieties rapidly increase. The water harbors and supports but a few kinds, all of which have many points of resemblance; while on the land, each zone and every locality has its peculiar birds. Only a very few birds literally inhabit all parts of the globe. Among these, so far as is known at present, there is not a single land bird. The Turnstone (strepsilas interpres), for instance, is found on the coasts in all parts of the world, because it finds the same conditions of life, as well in the eastern as in the western hemisphere.

As a rule, the circle of bird extension reaches farther in the longitudinal than in the latitudinal direction. In the northern parts of the globe, for example, are found many birds which abound more or less in all other parts in or near the same latitude; while a few hundred miles to the north or south may produce a very decided change in the number and kinds of birds. The flying capacity of birds is not coincident with the circle of extension. Good flyers may be confined to a comparatively small area, while indifferent ones may have the range of a much larger circle. The regular travels and migrations of birds do not enlarge their circles of extension.

SPECIES AND ORDERS.—So far as at present known, the number of all the different described and non-described species of birds may be estimated at about eight thousand. The order of Parrots numbers three hundred and fifty, birds of prey, four hundred; Pigeons, about three hundred; scratchers, three hundred; shortwingers, ten, stilts and swimming-birds, about six hundred kinds. The remaining species are distributed to other orders. America has a larger number of the different kinds of birds than any other grand division of the globe. Next follow in this regard, in the order named, Asia, Africa, Oceanica, and Europe.

The following remarks may be made respecting different orders of birds. The first order includes Parrots, wanting in Europe; the second order embraces Sparrows, the third order, Ravens; the fourth order, birds of prey; the fifth order, the spreading birds; the sixth order, singing birds, and the seventh order, climbing birds; the eighth order, Humming-birds, is confined to America; the ninth order, light birds, is chiefly found within the tropics; the tenth order, cooing birds, and the eleventh order, scratchers, are represented in all parts of the world; the twelfth order, short-wingers, belong to Africa, Oceanica, and more generally to America; the thirteenth order, stilts, and the different orders of swimming birds, have representatives in all portions of the globe.

Habitat and Trades.

THE HABITAT, or residence of birds, depends on the means it affords them for a living. Birds that live on the water ascend high mountains, and still higher rise the stilts, as they are less confined to the water. The forests are densely populated with birds of various kinds. Oceans, seas, and lakes support millions of individuals of the same or similar kinds. Hatching-time collects them in multitudes on single rocks and islands. Frequently, the greater the uniformity of the open or forest land, the greater seems the variety of the feathered tribes. The nearer an approach is made to the equator, the greater also is this variety, as the countries of the torrid zone present an increase of the varied conditions of life. The like wonderful variety, too, obtains in countries where forests and prairies, mountains and valleys, dry lands and swamps or water, alternate with each other. A river passing through a forest, a swamp bordered with large trees, or an inundated part of a forest, always collects great multitudes and varieties of birds; for where the productions of land and water are combined, the greater the variety and richness of the food. As the supply of food attracts birds to certain localities, so the want of it compels them to abandon others. They manifest great sagacity in selecting the best portions of certain districts. They pry into every hiding-place, into every crack, cranny, and hole, and pick up every digestible thing.

TRADES OF BIRDS.—Considering the different modes in which birds support life, it may be said that they have different trades or callings. Some, like Pigeons, and other granivorous birds, pick up the grains they find on the surface of the ground; others of the same kind strip the husks from the grain, and some scratch seeds and roots out of the ground. Fruit-eaters pluck fruits with their beaks; insect-eaters seize their prey on the ground, or on the leaves or branches of trees. Some of these latter birds labor very hard for a living, searching with their tongues for insects in the innermost recesses of their barking-places. Ravens pursue all these various trades, and operate on a small scale as regular robbers. Falcons and Hawks, or Eagles, are constant hunters. Vultures are scavengers, and other birds may be regarded as beggars and spongers.

Aquatic birds have their trades as well as land birds. Many of the former pick their food lying in plain sight; others ransack the hiding-places of other animals; some are omnivorous; others carnivorous; some draw their food out of turbid water or soft mud; others obtain theirs by diving deep in the water, and others, perceiving their food from a great height, dart thence upon their prey.

Genesis and Development.

THE EGG.—After impregnation of the female bird, one of the little yelk bodies which adhere in the ovaries, starts out from the rest, absorbs from the blood all the matter pertaining to the yelk, and is finally itself transformed into a yelk, growing to its proper size. Detaching itself, it slides into the ovi-director, which shows, during the time of laying, an increased activity, and secretes albumen. Both the yelk and albumen are now pushed forward by the contractions of the ovi-director, and arrive finally at the lower expansion of the same, or the so-called womb (uterus). Here they assume the regular egg form, and receive the shell skin and the calcareous shell. The latter is at first soft and adhesive, but soon hardens, and the egg is complete. By the contraction of the fibers of the muscles of the uterus, the egg is pushed forward into the cloaca, where it is probably colored, and whence it is finally ejected. The size and form of the egg, which may depend on the form of the uterus, vary greatly. There is supposed to be, in general, a certain ratio between the size of the egg and of the body producing it; but there are large birds that lay comparatively small eggs, and small birds that lay comparatively large eggs. The form of the bird's egg usually resembles that of the hen's egg, but some birds' eggs correspond more to the form of a top or pear, while others are equally rounded at both ends and are very oblong in shape. Eggs laid in hollows or crevices are generally white, or have only one color; those laid in open nests are, for the most part, speckled. The number of eggs laid by one bird varies from one to twenty-four, provided none are taken from the nest during the process of laying, as in that case the number may be increased ten or more. But the great majority of birds lay from four to six eggs.

INCUBATION.—Having completed the task of laying her eggs, the female bird enters upon the process of incubation or hatching, in which she is often relieved by the male. The process is accelerated by the warmth of the body of the female, which now possesses a high temperature. Some birds utilize the heat of the sun and of vegetable fermentation. The duration of the hatching process varies with different kinds of birds. The Ostrich takes from fifty-five to sixty days; the Humming-bird only ten or twelve, and birds, on an average, from sixteen to twenty-five days. For the formation and development of the chick or young bird, a temperature of about ninety-six degrees (Fahrenheit) is required. The requisite warmth may be obtained from the parent bird, or by artificial means, as breeding-ovens, etc. That famous story-teller, Pliny, relates that Julia Augusta, the wife of Tiberius, hatched eggs in her bosom. Besides warmth, the access of atmospheric air is absolutely necessary to the development of the chick. Without this, the egg will inevitably become addled.

PROGRESS OF DEVELOPMENT.—The effect of warmth may be seen after a few hours, and in hen's eggs, in about ten or twelve hours after the commencement of the incubation. The white round dot, or treddle, assumes a more oblong form, and its surrounding white rings are widened and increased in number. At the beginning of the second day, a small projection appears, and, in thirty hours, there is seen in its bladder-like cavity filled with a clear liquid, a turbid, cloudy jelly. Toward the end of the second day, the first traces of blood appear in reddish dots, streaks, and lines, gradually combining and forming a sort of net-work. These are the rudiments of the blood-vessels, which appear more distinct on the third day, forming branches at first, and finally a central point, eventually developing into the heart. Soon after its first manifestation, it begins to move; it expands and contracts: life has awakened, and manifests itself. Next are observed three dots like small bubbles, which originate the head, two of them being rudimental eyes. These are of a dark hue, the other dot is colorless. Now appears a row of little bubbles in pairs, which, taking a downward course, will combine into a backbone. Two developing plates mark out the circuit of the abdomen, and, at the same time, are exhibited traces of the stomach and the intestines.

On the fourth day, the yelk is enlarged, but loosened and thinned; the albumen is decreased, and the blood-vessels increased; the separation of arteries from veins is progressing; the head is bent down toward the posterior portion; the heart becomes more distinct in its manifestations; the vessels of the brain, and the rudiments of the mandibles, the wings, the legs, and the liver begin to be discerned, the last as a reddish gray jelly-like mass.

The fifth day presents the heart and intestines more fully developed, and the breast almost covered by a compound substance, beginning from the backbone and the wings. At the close of this day, the beginning of the lungs is observable; the heart is surrounded by a transparent serk, and the backbone is plainly shown.

On the sixth day, the egg-skin is formed, appearing like two closed bladders—the external one for the curds and the internal for the amnion. On the abdomen, enlarged by the admixture of albumen, the vessels are spread out; the several parts of the embryo are more distinctly developed, and it sometimes exhibits, at the close of this day, a kind of motion.

The embryo, on the seventh day, swims in the amniotic fluid; it is now about an inch long, and its head is almost as large as its body. The brain, appearing like a soft, slimy mass, presents to view its different parts, and the backbone shows traces of the beginnings of cartilaginous formations. The ribs are seen in white streaks. The throat, the crop, the gizzard, gall-bladder, and milt are likewise discernible.

The eighth day exhibits the chick increased in size; the breastbone begins to be formed, and white streaks, the beginnings of muscles, appear around the developing bones.

The ninth day discloses a small process on the head, forming the

upper mandible; the transparent eyelids become visible around the eyes; the heart, inclosed in its sack, pulsates twelve times in a minute; the brain becomes more compact, and its cartilaginous bony portions assume greater distinctness.

On the tenth and eleventh days, the embryo still increases in size; the large head is considerably lessened, and lies between the legs, almost covered by the wings; the gall-bladder is filled, and on the skin are seen small protuberances, whence feathers will spring in due time. On the two following days, the chick begins to move visibly, and attains a length of about two inches; downy feathers appear on the part of the body near the pelvis, and on the back, wings, and legs; the limbs are more fully developed, and the feet and toes begin to be covered with white scales; the bill presents itself in a cartilaginous form; the brain has almost attained its future size and shape; the cartilaginous brain cover (the skull) begins to ossify, and the lungs gain their proportionate size; the cartilaginous rings are traceable on the windpipe, and the urinary vessels in the kidneys, while the urinary tube, the ovaries, and the oviduct are easily distinguished. The muscles are soft and of a white color, the large sinews are more distinct, and the cartilaginous forms of the bones begin to ossify.

On the fifteenth and sixteenth days, the chick reaches almost its full size; the white scales on the feet and toes are converted into a horny casing; the larger wing-feathers begin to shoot out, and the embryo chick, if disturbed, opens and shuts its beak. During the seventeenth, eighteenth, and nineteenth days, the cutis is spread over the whole inner surface of the egg; the albumen becomes almost totally absent; and the yelk-bag collapses and passes through the navel ring into the abdominal cavity. The chick dons its feathery robe, lying inclosed in the amnio, folded in a compact form, and leaning sideways on its breast, with its head under the right wing, and its legs drawn up to the abdomen. It begins to be active now; it opens and shuts its bill, gasps for breath, and often utters a distinct "peep." The head is completely developed, and the brain has assumed its permanent form.

On the last two days, the yelk altogether disappears in the cavity of the abdomen, and the chick fills the entire egg. It breathes, peeps, and, if taken out of the egg, puts out its tongue. Several hours before its exit, the chick, on the twenty-first day, rubs the horny protuberance upon its bill against the shell of the egg. This begins to crack, little pieces fall off, the shell is torn asunder, and the chick, stretching out its new-found legs, pulls its head from under its wing, and comes forth from its prison-house to air, light, and freedom.

Youth and Maturity.

GROWTH.—A few birds only are as fully developed in the egg as chickens and ducks, and these only that are destined by nature to follow their parents and rise about as soon as they make their debut in the world. The birds that have, when full grown, the greatest mobility and power, are among the most helpless in infancy. These that run from the nest (Nest-runners), come into the world well-winged and with developed senses, and have a pleasing appearance, being already, to a certain extent, perfected. Others that attain similar perfection by slow degrees, and remain, as the remainder, in the nest (Nest-squatters), are ungraceful, and even deformed in appearance; their development requiring a longer or shorter time, according to their kind. The smaller kinds of nest-squatters are usually fledged in about three weeks; the larger often require several months. Most of them take years to become equal to their parents, for the youth of a bird can only be said to be extinct when it appears in the dress of its parents. Many of these birds have at first a feathery dress, which has no resemblance to that of the parents. A majority of these young birds are covered like the females, and the difference of sex can be known by the external appearance only when they have assumed the dress of maturity.

Eagles and several other birds are often three, four, or five years old before they can be called old birds, having donned their permanent plumage.

MOULTING.—The different changes to which the plumage is subject, are caused by the peeling off and discoloring of the feathers, or the process of moulting. The consequence of the peeling off is usually no embellishment, and the darkening produces, in certain portions of the plumage, a change of colors. The young White-headed or Bald Eagle, for instance, is covered almost wholly in a sober brown dress; in due time, the full-feathers become spotted, the white specks on them gradually becoming larger; at length these specks run into each other, till finally the whole tail turns white. Nearly the same transformation occurs in the head-feathers, except that with these the process begins in white streaks, gradually advanced, till the entire head is rendered perfectly white. The same discoloring is seen in the American Goldfinch, whose plumage is changed every spring from a light brown to a bright yellow, black and white dress, not, in the fall, changed back to the sober brown. Many other birds are subject to similar changes.

The moulting of birds usually takes place when the feathers have become damaged by long use, by the effect of light, dust, wet, etc. Moulting generally begins shortly after hatching. The unusual heat of the body, and other causes occurring at that period, injure the feathers. The change of feathers takes place at the same time on different parts of the body, but always on the corresponding parts of the two sides. On many birds, the smaller feathers only are replaced at the first moulting; the larger feathers of the wing and tail, at the second. Several years are required with some birds to replace the larger feathers, as only two new feathers, one in each wing, are formed at the same time. The moulting, in some cases, takes place so quickly that the birds, as Ducks for example, become totally incapable of flying. When a bird is in good condition, it appears, after each moulting, brighter colored, and this brightness of color is not diminished, but increased by age. Birds, at moulting time, are subject to many diseases, and, as this process is necessary to their well-being, anything that interrupts it endangers life.

LONGEVITY.—It may be laid down, as a general rule, that birds attain, comparatively speaking, a great age. Pigeons sometimes live twenty years; Canaries, in cages, often sixteen or eighteen; Parrots, sixty to one hundred; Geese, eighty to ninety; Linnets, and other small birds caged, often ten to fifteen years; Eagles, over one hundred and twenty, and Laurens, as there is ample testimony to prove, live from two hundred to two hundred and fifty years.

DISEASE AND DEATH.—Birds, generally speaking, are not very liable to disease. But they are exposed to many accidents and to many enemies. They are continually destroyed by birds of prey and other animals. Sometimes epidemics have been known to prevail among certain species of birds, by which great numbers have been swept off. At first view, it seems rather singular that the body of a bird that has died a natural death, is seldom found. But investigation has disclosed the fact that birds generally have an anticipation of their approaching dissolution, and this feeling prompts them to seek concealment. Birds that would otherwise never enter any such place, prompted by this feeling, conceal themselves in the cavities and hollows of trees, and creep under stones and into the thick tussocks of grass, or other like places of concealment. Only now and then is the dead body of a bird found under its resting-place. The dead bodies of aquatic birds are sometimes washed ashore, without bearing any traces of death by violence.

Subsistence, Mating, and Breeding.

VIGILANCE AND ACTIVITY.—Birds are active and lively. From break of day till late at night, and often during the whole night,

birds are awake and on the alert. The longest days seem too short for them, and the shortest nights not short enough. At early morn we hear in the woods the voices of birds; the same voices are heard during the day and long after nightfall. Birds seem to need very little time for rest or sleep; three or four hours appear to suffice. From this rule, birds of prey and Vultures are exceptions; as, among birds, they are the indolent classes.

FEEDING.—All birds gifted with the faculty of singing, hail the approaching day with a matin song, especially in the time of mating or breeding. After this they take breakfast. Most birds take two regular meals daily—one in the morning and the other late in the afternoon. Midday is usually devoted to rest, or to putting in order and cleaning their feathers. Such birds as depend on luck and chance for their supply of food, as birds of prey, generally put up with one meal per day. Vultures have sometimes to content themselves with a single meal every other day, as they never search for food, but depend on what chance and good luck may offer.

Birds, in general, to use a common expression, live only "from hand to mouth." They eat food as they find it, unconcerned about the morrow. Our Woodpeckers, however, generally store away a considerable amount of food for winter use. They deposit it in different places, as the hollows of trees, cracks in their bark, and holes in the ground. Some other birds, as the Nuthatches, collect food and hoard it in different storing-places.

BATHING AND RESTING.—Birds drink immediately after feeding, and then take a bath. This is taken in water, or in dry sand or dust, and sometimes, in winter, in snow. Water-bathers exercise a good deal after their bath, by flying about and shaking themselves. Then, after a short rest, their plumage is put in order and preparation made for a foraging excursion. If the expedition proves favorable, then, toward evening, the birds betake themselves to certain well-known haunts, for a social interview with kindred and friends, before retiring for the night. The singing-bird, ere he goes to his night's repose, warbles his sweetest songs, and then retires, either in company with others of his kind, or, in breeding time, with his little sitting mate or with his young ones. He never goes to sleep without having previously engaged in some sort of conversation, carried on by means of various gestures and chatterings. He ceases only when overcome by fatigue.

MATING.—The breeding season with birds is always in the spring. In the torrid zone, it occurs in the rainy season, which perhaps corresponds to our spring. Birds generally practice monogamy; very few among them are polygamists. A pair once united, are, as a general rule, mated for life, and it is only in exceptional cases that either proves conjugally unfaithful. Yet, as there are more males than females among birds, it naturally happens that a number of male birds, including those that have lost their female companions, are compelled to remain unmated. These are described as restless, wandering from place to place in search of mates, and resorting to various tricks and devices to entice away females from their allegiance to their "liege lords." The result is, that the latter often become furiously jealous, and quarrels, and even severe fights, frequently ensue. Though the attempted seduction is sometimes successful, yet it often happens that the female sallies forth with her mate to drive off to a respectful distance the impudent intruder on their domestic felicity. So much for constancy in female bird mates. On the other hand, it has been observed that a female bird having lost her mate, became, in half an hour afterward, the partner of another male; and this second mate being killed, she did not hesitate a moment to listen to the addresses of a third wooer. The male birds are ardent in their courtship, displaying all their beauty and attractions. Some woo by gentle longing calls and by singing; others, by a sort of dancing or gamboling in the air; sometimes the male pursues the female for hours,

Part 3.

Price, $1.00.

THE BIRDS OF
NORTH AMERICA

DRAWN FROM LIFE

AND

Uniformly Reduced to One-Quarter their Natural Size

BY

THEODORE JASPER, A.M., M.D

Jacob H. Studer, Publisher

COLUMBUS, OHIO

ROBERT CLARKE & CO.
63 WEST FOURTH STREET
CINCINNATI, O

C. D. CAZENOVE
13 BEAUFORT BUILDINGS, STRAND
LONDON

settled parts of the country he is rarely or **never found**, but seems to prefer the company of men. His nest is built in briers or blackberry bushes, and is composed of thin branches and **roots**, stuck together with mud, lined inside with hair and finer fibers. The female lays five eggs, of a bluish tint. He leaves in September to winter in warmer latitudes.

The Maryland Yellow Throat. (*Geothlypis Trichas.*)

Fig. 3, Male. Fig. 4, Female.

This neat little bird inhabits chiefly such briers, brambles, and bushes as grow luxuriantly in low, watery places, his business and ambition seldom leading him higher than to the tops of the underwood, and he might properly be denominated "Humility." Insects and their larvæ are his principal food. He dives into the thicket, rambles among the roots, searching around the stems, examining both sides of the leaves, raising himself on his legs to peep into every crevice, and amuses himself with a simple, but not at all disagreeable twitter, "whit-ti-tee! whit-ti-tee!" which he repeats in quick succession, pausing, now and then, for half a minute. He inhabits the States from Maine to Florida, and westward to the Mississippi. He is by no means shy, but unsuspicious and deliberate. He often visits the fields of growing rye, wheat, or barley, and is of much service to the farmer by ridding the stalks of vermin that might destroy his fields. He lives in obscurity and peace, and seldom comes near the farmhouse or the city.

He builds his nest about the middle of May, in the midst of a thicket of briers, among the dry leaves on the ground. Sometimes it is arched over, and but a small hole left for entrance. It consists of dry leaves and fine grass, lined with coarse hair, etc. The female lays five eggs, semi-transparent, marked with specks of brown and reddish brown. The young leave the nest in the latter part of June, and a second brood is sometimes raised in the same season. They return to the South early in September.

PLATE VIII.

The Wood Duck, or Summer Duck. (*Aix Sponsa.*)

Fig. 1, Male. Fig. 2, Female.

This is the finest of all our Ducks, and the beauty of its dress is in perfect harmony with its gentle manners. A characteristic trait is the moving of its tail from one side to the other, which sometimes looks almost like wagging. It swims with as much ease and grace, and seemingly with as little effort, as it flies among the branches and trunks of trees. The cry of the female is a long-necked "Phek-wee-woo-wee!" and the warning sound of the male a not less melodious "Oeek! O-eek!" It seems to stun the neighborhood of men less than any other Wild Duck, and is by no means in a hurry to leave its breeding-places, even if buildings are in construction close by. Easier than the rest of the tribe, the Wood Duck gets domesticated to, and regularly breed in captivity, if a suitable chance is offered them.

They live mostly on grain, several aquatic plants, chestnuts, acorns, beech-nuts, etc., also on worms, snails, and other insects, which they pick up among the dry leaves or catch in the air. Their tall beauty and loveliness shows itself best chiefly before and during mating time. Toward March the flock separates, and every single pair now looks out for a convenient breeding-place. To this end the male roams about the woods, alights on a high tree in which he expects to find a hole for a nest, walks easily on its limbs, inspecting every hole for one find, and is often perfectly satisfied with a hole made by the fox squirrel, or even a cleft in a rock. The female appears herself with astonishing ease through the entrance,

which often seems to be a great deal too narrow for her. The male keeps watch outside during inspection by the female, encouraging her by his tender chatterings, or warning her of supposed danger by his timely "O-eek! O-eek!" after which both quickly take to flight. If they have once built a nest they return to it every year.

The male, although very peaceful, is very courageous when his jealousy is aroused. After other male coming near him is always kept at a proper distance by unmistakable signs and motions. The female begins to lay in the first days of May. The eggs, seven to twelve in number, are small, oblong, and perfectly white. The hatching-time lasts, as with most of the Duck tribe, twenty-seven or twenty-eight days. When the last egg is laid, the female lines the nest with the soft down of her breast, and covers the eggs with the same when she flies out. While she takes all the parental cares to herself, the male repairs to a suitable water place to pass through his moulting time, which begins in July, and is ended in the first part of September, giving him a dress distinguished from that of the female only by the white marking of his throat and the greater brilliancy of his plumage.

The nest of the Wood Duck is sometimes at a considerable distance from any water, and quite high from the ground. From the entrance to the nest itself, it is sometimes over six feet. As soon as the young ones are all hatched, the female carries them, one by one, in her bill, to the water, leaving them to the care of the male, till she has brought the last one, often she herself takes care of them again. If the tree on which the nest is, happens to overhang the water, she merely tumbles them out of the nest. Wood Ducks generally live together in small flocks of from six to twelve—occasionally they are seen in flocks of more than a hundred; this occurs chiefly in the fall. Toward October the young ones begin to moult; at the same time the male parent, who reappears now in his bridal dress, joins them again. The flesh of the Wood Duck is very tender and in good esteem.

The Short-tailed Tern. (*Hydroche-lidon Plumbea.*)

Fig. 3.

This bird is often observed in fresh-water marshes, in flocks numbering from four to ten; it is seldom seen in salt-water marshes. Its flight is very graceful. Its food consists of grasshoppers and insects generally, which it picks up, while on the wing, from grasses or rushes, as well as from the surface of the water. It frequently associates with

The Black Tern. (*Hydroche-lidon Nigra*)

Fig. 4.

The Black Tern is a little less in size than the preceding, which it resembles in every respect. They are found on fresh-water marshes, mill-ponds, etc., and are most numerous on the marshes of the Mississippi and its tributaries. Their nests are very artlessly constructed, in large mounds of rank grass, and contain each four eggs, of a greenish buff color, spotted with amber and black, chiefly at the larger end. The young ones of the first season (Fig. 5) have the head white, and the neck and breast irregularly spotted with black and white.

It was found, on dissecting these birds, that they feed exclusively on insects, their stomachs never containing any small fish.

Mr. Audubon, in his valuable work on "Birds of North America," writes as follows of this bird:

"The Black Tern begins to arrive from the Mexican territories over the waters of the Western country about the middle of April, and continues to pass for about a month. At that season I have observed it ascending the Mississippi from New Orleans to the head waters of the Ohio, then crossing over the land, and arriving at the Great Lakes, beyond which many proceed still farther northward. But I have rarely met with them along our Atlantic shores until autumn, when the young, which, like those of all other Terns with which I am acquainted, mostly keep by themselves until spring, make their appearance there. Nor did I see a single individual when on my way to Labrador, or during my visit to that country. Often have I watched their graceful, light, and rapid flight, as they advanced and passed over in groups of twenty, thirty, or more, during the month of May, when nature, opening her stores anew, benignly smiled upon the favored land."

PLATE IX.

The Woodcock. (*Rusticola Minor.*)

This bird, is generally known to our sportsmen, is represented at the bottom of the plate. It arrives in the United States in March, and if the season is mild, even earlier, and stays till the first frosts forebode the approach of winter. It is sometimes noted here in December, and it may be that in mild seasons some of these birds remain until spring. During the day the Woodcocks keep to the moist, or wooded swamps and thickets; toward evening they usually fly out to the broad open glades, which lead through the woods, or to meadows and various places in the neighborhood. A carefully hidden observer may see the Woodcock pushing his long bill under the decayed leaves and turning them over, or feeling his toes close to another as the damp soft ground, as deep as his toes. Such his bill will permit, to get at the larvæ, bugs, or various hidden beneath. In a similar manner he examines the fresh cow-dung, which is soon populated by a vast mass of larvæ of insects. He never tarries long in any place. Larvæ of all kinds of insects and naked worms, expressible any worm, the form his principal food.

If so fed answers his favorite resorts in watery recesses inland are generally dried up, or descends to the marshy shores of our large rivers.

The female Woodcock usually begins to lay in April: the nest is built of a quantity of the moist, frequently at the roots of an old stump; it is constructed of a few withered leaves and stalks of grass put together with but little art. The eggs are four or five in number, about an inch and a half long, and about an inch in diameter, tapering suddenly at the small end; they are of a dun clay color, thickly marked with brown spots—particularly at the large end the spots are interspersed with others of a very pale purple. The young Woodcocks, when six to ten days old, are covered with down of a brownish white color, and are marked from the bill along the crown to the hind head with a broad stripe of deep brown; another line of the same color curves under the eyes; and runs to the hind head; another stripe reaches from the neck to the hindmost of the bill, and still another extends along the sides under the wings. The throat and breast are considerably tinged with yellow, and the quills of the eye are just bursting from their light blue sheaths, and appear marked as on the old birds. When taken they utter a long, clear, but very feeble "peep," but looks less than that of a mouse. They are, on the whole, far inferior to young Partridges in running and walking.

The Woodcock is a nocturnal bird, seldom stirring about before sunset, but at that time, as well as in early morning, especially in spring, he rises by a kind of spiral course to great heights, uttering now and then a sudden squeak; having gained his utmost height he hovers around in a wild irregular manner, producing a sort of murmuring sound, and descends with rapidity in the same way he arose.

The large head of the Woodcock is of a very singular conformation, somewhat triangular, and the eyes set at a great distance from the bill, and high up in the head; by this means he has a great range of vision. His flight is slow; when flushed at any time he rises to the height of the bushes or the underwood, and usually drops down again at a short distance, running off a few yards as soon as he touches the ground.

The Wood Thrush. (*Turdus Mustelinus.*)

Fig. 1, Male.   Fig. 2, Female.

The Wood Thrush is one of our best and sweetest singers. Audubon writes of him as follows:

"The song of the Wood Thrush, although composed of but few notes, is so powerful, distinct, clear, and mellow, that it is impossible for any person to hear it without being struck by the effect which it produces on the mind. I do not know to what instrumental sounds I can compare these notes, for I really know none so melodious and harmonical. They gradually rise in strength, and then fall in gentle cadences, becoming at length so low as to be scarcely audible, like the emotions of the lover who, at one moment exults in the hope of possessing the object of his affections, and the next pauses in suspense, doubtful of the result of all his efforts to please.

"Several of these birds seem to challenge each other from different portions of the forest, particularly toward evening; and at that time nearly all the other songsters being about to retire to rest, the notes of the Wood Thrush are doubly pleasing. One would think that each individual is anxious to excel his distant rival, and I have frequently thought that on such occasions their songs were more than usually effective, as it then exhibits a degree of skillful modulation quite beyond my power to describe. These concerts are generally continued some time after sunset, and take place in the month of June, when the females are sitting."

The Wood Thrush inhabits almost the whole continent of North America, from Hudson's Bay to the Gulf. The very next morning after his arrival he will mount to the top of some small tree and announce himself by his sweet song, which, although not containing a great variety of notes, is exceedingly soft and melodious, poured forth in a kind of ecstasy, and becoming more intensely at every repetition, especially if several of them be heard at the same time, in different parts of the wood, each trying to outdo the other. He is always in good humor, and his voice is often heard as many days, from morning to nightfall. His favorite retreats are thickly shaded hollows, trees grown rich shrubbery and creeks or rills, searching with alder bushes and wild grapes. It is in such places, or near them, that he builds his nest, a little above the ground. It is constructed externally of withered leaves or prairie grass; on some are layers of knotty stalks of withered grass mixed with mud and smoothly formed; the inside being composed of the dry leaves of plants. The female lays four, sometimes five, light blue eggs. The Wood Thrush is a shy and unobtrusive bird, appearing rather timid or in places, and feeding on different kinds of berries, as well as on beetles or caterpillars.

On his migration to the South he never appears in the open plains, but hops and flies swiftly through the woods. Occasionally he takes a rest on a low branch, uttering a low chuckling sound, and jerking his tail up and down at each note; then for a few moments he keeps perfectly still, with the feathers of his neck and back a little raised.

The Yellow-bellied Woodpecker. (*Picus Varius.*)

Fig. 3, Male.   Fig. 4, Female.

This is one of our resident birds, and is often to be met with in the thickets of the woods in midwinter. It is generally considered a handsome bird, and in its manners and mode of living resembles the small spotted Woodpeckers.

He is frequently seen in their company, especially in the fall, when visiting the orchards. Its nest is usually in a dry old tree, or in a large fallen branch, the entrance to which is small for the size of the bird, and passing down in a slanting direction it expands toward the place where the eggs lay, which are from three to four in number and of a pure white color. Nests containing eggs are invariably to be found from about the middle of May to the first of June. This bird is seen with almost everywhere, but not in great numbers, from Hudson's Bay to the Gulf of Mexico. Its food, like that of all the Woodpeckers, consists chiefly of insects and their larvæ, and to some extent of berries.

The Scarlet Tanager. *(Pyranga Rubra.)*

Fig. 5.

This beautiful bird is an ornament to our woods. It is almost destitute of song, being endowed with a few notes only, which resemble those of the Baltimore Oriole. It may be found in all parts of the United States, even as far up north as Canada. It rarely visits the habitations of man, but frequently orchards, where it sometimes settles down on an apple or pear tree. Its nest, which it builds in the middle of May, on a horizontal branch, consists of stalks of broken flax and other dry fibrous matter loosely woven together. The eggs, three or four in number, are of a dull bluish color, spotted with brownish purple.

It seems not to be very shy, but allows you to approach it very near, and is frequently sitting right above your head while you are looking for it in the distance, misled by its notes, "chip, cheer," which seem to come from a great distance.

The female is green above and yellow below; the wings and tail brownish black, edged with green. The male has a spring and a summer dress. Our plate shows him in the spring dress. This changes, soon after the young are hatched, into one similar to that of the females—green above and yellow below; and in the time between this and his bridal dress, he is often speckled with red, which is produced by the red points of the feathers; for, with the exception of the points, these feathers are of a bluish and sometimes a yellowish white; but they lie so regularly on the living bird that the white parts are invisible.

PLATE X.

The Snow Owl. *(Nyctea Nivea.)*

Fig. 4.

The Snow Owl, the largest of all the so-called Day Owls, inhabits all parts of the North. However near men have approached to the pole, they have seen this Owl, not only on the land, but they have observed him likewise sitting on icebergs, or flying close over the water with powerful flapping of the wings. It is, therefore, probable that they inhabit not only the whole of North America, but also the corresponding latitudes of Europe and Asia.

In extremely cold winters they regularly wander southward, and are by no means scarce in Illinois. Several of them were shot near Chicago, in the winter of 1871-72. Our drawing was prepared from a beautiful female specimen.

A gentleman from Cuba assures us that he has frequently seen this Owl there.

Some ornithologists of Europe hold that the color and markings of this species are different at different ages, and that some are like the one on our plate, while others are almost or perfectly white. It may be so; but on dissection the white ones have been invariably found to be males and the others to be females. The white Owls are the smaller.

During the summer they generally keep in the mountainous part of the North; in winter they take up their abode in the plains. In his manners, the Snow Owl has many peculiarities. In his quiet sitting position, his resembles all other large Owls; but his movements are quicker and more graceful, his flight being like that of the slow-flying birds of prey. In boldness and tenacity he surpasses all the rest of the Owl tribe. His food consists chiefly of small quadrupeds, such as the muskrat; partly also of fish, which he catches with great skill, in nearly the same manner as the Fish-hawk, sitting on a projecting rock and watching for them, until they come to the surface of the water. In winter he prefers the evening or the night to day-time for hunting. His cry is a rough, harsh "craw! craw!"

The eggs are laid in the month of June. Their number varies from five to ten—a remarkable number for a large bird of prey like the Snow Owl; they are oblong and of a dirty white color. The nest consists of a small cavity in the ground, lined with withered grass and a few feathers from the mother bird. Both parents are much attached to the young, and on the approach of man, the female flies off a short distance from the nest, and, feigning lameness, remains with spread wings, lying on the ground, in order to coax the enemy away from the nest. It has been tried many times to keep Snow Owls in cages; but they invariably died in a short time without any apparent cause.

The Snow Bunting. *(Plectrophanes Nivalis.)*

Fig. 6.

The Snow Bunting inhabits, like the Snow Owl, the northern regions not only of this continent, but also of Europe and Asia. His home is in the mountains, where he builds his nest in crevices of rocks or under stones; the outside of it is composed of dry grass, moss and lichen, the inside of feathers and soft down; the eggs are five to six in number, are so irregularly marked and colored that a description of them is almost impossible. The song of the male is very pleasant but short. The young birds, when fully fledged, remain for a short time in their old haunts, then form large flocks and begin their regular wanderings. As hardly any other birds fly in as large flocks, at least not in northern regions, their wanderings attract the attention, not only of naturalists, but of almost everybody. In Indiana they appear only in small groups of from sixteen to fifty. They travel also considerable distances over the sea.

In their manners, Snow Buntings resemble Larks. They fly easily, with little flapping of the wings, in long curving lines, generally at considerable heights, and sometimes just above the ground. They are of a lively, troublesome disposition, and seem to be in good humor even on the coldest winter days. In summer they subsist chiefly on insects; in winter they feed also on several kinds of seeds. It is very amusing to see a flock of them in winter, on the snow-covered fields, on a foraging tour. They hover over the ground, a part of them alighting to pick up what little seed they can find on such withered plants as extend above the snow, the rest flying just over them a little further along, and then alighting also; after a while the first party fly over the others, and in this way they go over the whole field. Their cry on such occasions sounds like "tsi," sometimes it is a shrill "twirr," uttered during the flight. Our plate represents this bird in its winter dress. The summer dress of the old male is really handsome, notwithstanding its plain colors. The whole middle of the back, the tips of the primaries, and the middle of the tail feathers are black. There is also a black spot on the metacarpus. All the rest of the plumage is snow white.

PLATE XI.

The Yellow-shanked Snipe. (Gambetta—Scolopax—Flavipes.)

Fig. 1.

The Yellow-shanked Snipes arrive in the Northwestern States between the middle of April and the early part of May, on their way to the North, where they breed; and return as early as the latter part of August, or the beginning of September, making only a short stay. All the birds of this genus seem only to go northward to breed, and to return southward as soon as the young are able to fly. Single ones are to be met with in summer, or at almost any season; but as all of them are male birds, it is to be presumed they are either old bachelors or widowers, who can not bear to see the happiness of those who are mated, and therefore wander off toward the sunny South. There is more dignity in the manners and habits of the Snipes than in those of the Sand-pipers. Their flight is easy, and when they alight they flap their wings, and before laying them together, stretch them straight up, so that the tips touch each other. In case of need they swim and dive tolerably well. Their chief resorts seem to be the sea-coast and salt-marshes, as well as the muddy flats at low water, where they delight to wade in the mud; but it is rather the abundance of food they find there than the mud, that attracts them. They live on insects and all kinds of larvæ. You may sometimes meet with single ones, which show no shyness at all; but when in flocks they shun the gunner carefully and seem to distinguish him from less dangerous persons. It may be on account of these qualities that numbers of different kinds of Sand-pipers are found in their company, and seem to follow them as their leaders with great confidence. As a delicacy for the table, they are held in high esteem.

The Semi-palmated Sand-pipers. (Tringa—Actitis Semi-palmata.)

Fig. 2.

The principal places which these next little birds inhabit, are the seashores. Their legs are rather short in proportion to the size of the bird. They inhabit the same food as the Yellow-shanks. These birds inhabit almost every part of the North American continent. They migrate North in the spring, and should the season be open, remain quiet late in autumn, when they depart for their winter quarters at the South. They congregate in large flocks on the beaches and sand-bars, and meadows, along the sea-coast and on the shores of the interior lakes and streams. When feeding, they scatter about in search, perhaps when surprised, they run with a rapid movement, collecting in such close bodies that as many as twenty, and sometimes more, are killed at a single shot. When closely pursued, they rise off to one man offering a chirping note. If too near be reduced, they will shortly obey the call. They breed in the far North, the female laying four or five white eggs, spotted and blotched with black.

On their annual trip southward they sometimes penetrate far inland, following the softly and muddy banks of waters. In swimming they constantly move their heads backward and forward like Ducks.

A heavy down under the feathers of the breast makes them appear round and plump. In the fall the male and female are marked exactly alike.

The Great Tern, or Sea Swallow. (Sterna Hirundo.)

Fig. 3.

The Sea Swallows inhabit the northern parts of the temperate zones. They are found in great numbers on the North American lakes. In their wanderings they fly in a continuous current from one sheet of water to another, following, when it is possible, the course of rivers, and occasionally coming down to feed or rest. Their voice sounds like "kreee," and when frightened, like "kirk" or "krikk." Their food consists of small minnows, young frogs, tadpoles, worms, crickets, etc. They catch their prey when it is in the water by suddenly plunging down upon it; when they find it on the ground, they pick it up while on the wing. They build their nests on low islands, on the shores of rivers, or the coast generally, but not on sandy ground. They make small holes, or use such as they happen to find, for their nests, without lining them. The eggs are laid about the last of May, and are of a light yellowish brown color, speckled with purplish, reddish, and dark brown round or oblong spots. The female sits on them during the night, and the male occasionally in the daytime. During the warm sunshine the eggs are left uncovered. The young, which are hatched in about sixteen or seventeen days, soon leave the nest hiding themselves, in case of danger, among the pebbles, and only betraying their presence by their mournful peeping, when the parents are shot. The upper part of these birds is covered with a grayish mixture, and on the lower part the down is white.

They also stretch their heads toward water when sitting on the nest. Their flight is extremely graceful.

The young grow rapidly, and when only three weeks old are able to follow their parents.

PLATE XII.

The Baltimore Oriole. (Oriolus Icterus—Baltimore.)

Fig. 1, Male; Fig. 2, Female.

The Baltimore Oriole inhabits North America as far as the fifty-fifth degree of latitude. It is chiefly found in the vicinity of rivers, and seems to prefer a hilly country. It is only a summer visitor to the Northern States, where it makes its appearance in pairs, during the latter part of April or the beginning of May. It commences at once to build its nest, the material and construction of which vary according to climate and circumstances. In the Southern States, it consists of a Spanish moss, put together so loosely that for air can pass through it, the nest itself, and is always placed on the north side of a tree. In the Northern and Western States, it is hung on such twigs as are most exposed to the rays of the sun, and lined with the warmest and finest material. The bird, in constructing its nest, ties the material to the twigs with its bill and claws, weaving it strongly together, and giving the whole the respect a hanging bag, as shown on the plate.

In commencing its nest, he makes use of any material he deems suitable. A lady in Connecticut, while sitting at an open window, engaged in sewing, was called away for a few moments. A Baltimore Oriole, in the meantime, entered the window, and carried off her thread and several yards of small tape to the nest he was then building. The lady repeated the mischievous bird, and, mounting to the nest, found him winding it to his tope. Thus she succeeded in recovering, but the silk thread was so perfectly wound in that it could not be disentangled.

The female lays four and sometimes five or six eggs, of a light ash color and marked with dark spots, dots, and lines. The young are hatched in a fortnight, and in three weeks more are fledged. Before they fly out they often hang or climb among the nest like Woodpeckers. They are fed by both parents for a couple of weeks, and then left to take care of themselves. The food of the Baltimore Oriole consists of mulberries, cherries, and small fruit. In the spring they chiefly subsist on insects, which they pick up on leaves and branches of every thing. Toward fall they commence their return southward, flying high in the air, and always in the daytime. They generally fly singly with loud cries, and apparently in great haste. At sunset they alight to warble.

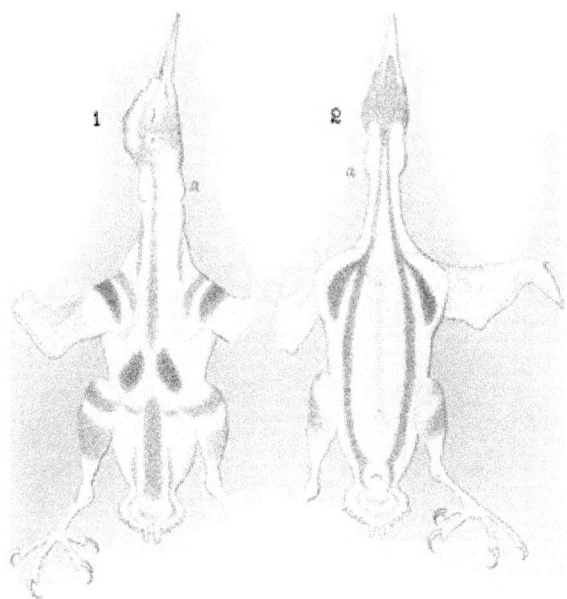

addressing her apparently in angry tones; but in general, the wooing is agreeable and soon brought to a successful issue, the female thence forward devoting herself to her mate with all possible affection and tenderness.

NEST BUILDING.—As soon as birds have mated, they look out for suitable nesting-places, unless they belong to the class of birds that have regular domiciles, to which they return from year to year. The nests are usually built in the central portion of the district which the birds inhabit, and are different in their architecture and materials. Birds of prey build nests on lofty trees or high rocks; the running birds generally build on the ground; others, on the branches, or in the hollows of trees (sometimes excavated by themselves), in the forks of bushes, or on the ground, among mosses or weeds. Aquatic birds make nests on the shore among rushes, weeds, or grass, or in sand-banks. Sometimes they fasten the nest to the rushes and leave it to float on the surface of the water, as do the Gallinules where the water alternately rises and falls. Some sea-birds build nests in rocky caves, or like the Puffins, dig out hollows in rocks for that purpose.

The concealment of the nest seems to be a great object with all birds. Where it is built in open places, it is so constructed as not to be easily observable. Some birds hardly make a nest at all, but lay their eggs on the bare ground, or on the naked surface of a rock. Some only make a small cavity in which to deposit their eggs; others fill the cavity with some soft and warm material; and others still, including those that build their nests on trees or form running nests on the ground, construct a sort of trestle-work as a basis for the nest.

The architecture of nests on trees and shrubs is as varied as are the birds that build them. Some nests are put loosely together; others are made in a more regular form out of tree branches, woody fibers, etc.; while other regularly formed nests are neatly lined with fine roots like threads and with hair or feathers. Some nests are roofed or arched over, and the entrance into others is lengthened to a kind of tube through which the birds creep in and out. The weaver-birds knot and sew their nests with long fibers or threads, and use, besides, little twigs and the soft wool-like material of different plants. Nuthatches are also adept in the art of nest building, forming the walls of their nests with loam, which made into mortar or paste by the saliva of the bird, becomes hard and durable. The principal use of the nest is to serve as a depository for the eggs and a cradle for the young. Some birds, however, build nests for their own amusement or for shelter in winter. To this class belong several kinds of weaver-birds, the Atlas and Collar birds, and Woodpeckers also, which always sleep at night in the hollows of trees, together with the domestic Sparrow, that regularly retires at night to a warm and well-lined nest, at least in winter.

The female bird usually constructs the nest, while her mate brings the materials. But among the weaver-birds, the nest is chiefly built by the male. Among other birds, the male acts as a sort of architect or superintendent, watching and directing the female during the progress of the work. The males of polygamous birds take no part in the construction of nests. Male birds generally, during the time of nest-building, entertain their female companions with sweet songs, or encourage them with their agreeable chatterings. Some birds' nests are made for a common use; different females lay their eggs in them together, which are often hatched alternately. Other birds build a large nest in common, and divide it into several apartments, each of which is used by a separate family.

As soon as the female bird begins to lay eggs, the temperature of her body begins to increase, and she soon has the so-called breeding-fever. This causes her to lose the feathers on some parts of her body. The process of hatching the eggs is chiefly performed by the female. She usually sits from the afternoon of one day till the forenoon of the next, and is then only relieved by the

male while she is successively absent, searching for food. Some birds, however, divide the labor of incubation, the male and female sitting alternately. The male Ostrich does all the sitting. The eggs, after the bird begins to sit, are turned almost daily, and are always covered with a soft down when both birds are absent.

CARE OF THE YOUNG.—The parent birds render no assistance to their young to facilitate their coming out of the shell. But as soon as they have effected their entrance into the world, they are cared for by their parents with assiduity and tenderness. At first they are fed with the tenderest food; then on that which is more substantial, and, as soon as they are able to leave the nest, they are taught to procure their own food and take care of themselves. All birds manifest great love for their offspring. They will protect them from peril and resort to all sorts of tricks to avert danger or turn away an enemy, risking their own lives in defense of their progeny. These, in turn, are in like degree attached to their parents, and listen obediently to their every call.

Migratory Birds.

MIGRATION.—As soon as their young can be safely left to themselves, several kinds of birds commence their journeyings to other countries. This journeying, when it occurs regularly every year, at a time certain, and in an appointed direction, is called "passage." But it takes the name of "wandering," when the traveling is the result of necessity, and therefore takes place neither at a certain time nor in a certain direction, ceasing when the cause that produced it is removed. It is also called "strolling" or "rambling," when the journey is confined within narrow limits, and is merely the result of a desire for a change of residence or for the sake of more abundant food.

Singing-birds make their passages or migrations every fall, and return in the spring. Aquatic birds set out on their passages long before the severely cold weather sets in. A majority of the birds of North America, and of Northern Europe and Northern Asia, migrate in a more or less southern direction; those in the Eastern hemisphere journeying northwesterly, and those in the Western hemisphere eastwardly, according to the prospect for finding plenty of food and a good winter-harbor. Rivers, valleys, and mountains serve for guidance to migratory birds. Sometimes these birds fly in pairs and sometimes in large flocks. The shy and feeble travel by night, the bold and strong both day and night. Before they set out, they grow restless and wander about, as if prompted to travel by an irresistible impulse. Birds taken from the nest when young, and kept confined in cages, manifest this same restlessness when the migrating season arrives.

It is a curious fact, that the birds that leave us the latest in the year, are the first to return, and those that leave the earliest, are the last to return. Birds that leave in November, return in February. North American birds travel to the southern parts of the United States or to Central America. Several kinds of European birds winter in the southern part of that continent, and many North African kinds, dwelling between the thirty-seventh and twenty-fourth parallels of north latitude, migrate south to regions within the torrid zone. Other winter-quarters for migratory birds are India, including Siam and South China. Birds in South America take a northern direction to Southern and Middle Brazil; and South Australian birds fly to the northern part of the island, some of them even to New Guinea and the neighboring islands.

All birds of passage, before they start, hold regular meetings, remaining in session for several days. They call in all those that pass by; and as soon as the flock has become large enough, the meeting is broken up, and the long flight begins. During the progress of the meeting, regular matters are held; leaders are chosen, and such birds as are judged incapable of remaining a long time on the wing and of keeping up with their companions are ejected.

Some naturalists even affirm that birds supposed to be too feeble to endure the tedious passage are put to death. The migratory flock flies in two rows or lines, so formed as to be gradually approaching nearer each other, and both ending in the one point, so as to form a figure resembling a harrow or the letter V. Some fly in direct lines, and some in disorder. Generally, those that fly very high sometimes take suddenly a downward course, fly low for a short time, and then rise to the same height as before. Some weaker kinds of birds fly only in the daytime, and as much as possible from tree to tree and from forest to forest, using the trees for shelter at night and during high winds and storms. Runners that fly with great difficulty, make rapid progress, alternately running and flying. Aquatic birds sometimes take to the water and swim, when they come upon a river or other body of water which they can use for helping them onward in their migration. The progress of migratory birds is aided by favorable winds, and retarded, often for several days, by contrary winds. The restlessness and excitement that birds experience before and during their flight, cease soon after they arrive at their destination.

The Wanderings of birds resemble their migrations, inasmuch as the former, with more or less regularity, take place at certain times. Wandering birds usually live in the higher northern latitudes, and are forced to migrate by having their feeding-places suddenly covered up by heavy snow-falls. Other wanderers fly from north to south within a certain zone, during very severe winters. Birds inhabiting the higher mountainous regions annually migrate to lower places, and, with the opening of spring, return to their former homes.

The Strolling of birds occurs at all seasons, and in all parts of the globe. Birds having no mates, go in search of them, and birds of prey are always strolling for food. Some birds seem to enjoy a strolling, irregular life, as they wander about without any apparent object. But wherever a bird may stroll, however long or short its journeyings, and however long or short the time of its sojourn abroad, its home is in the locality where it builds its nest and rears its young.

The Utility and Protection of Birds.

The Utility of birds to mankind is beyond question. They are our guardians against insects, whose ravages would, were it not for birds, lay waste the entire globe. Birds are held in high esteem even among savages. Their handsome forms, their bright and beautiful plumage, the astonishing celerity of their motions, their long, rapid, and lofty flights, and, above all, the delightful harmony and sweetness of many of their songs, win the admiration and affection of all human hearts. From the earliest periods of history, birds have always been the favorites of man, and especially of womankind. From them, more than from all other animals, we make selections for our entertainment, and for intimate and constant companionship in our private rooms, and even in our parlors. Many of our domestic and other birds supply us with feathers for use and ornament, while their eggs and flesh furnish wholesome and delicious food.

The Protection of birds has been often, in civilized countries, made the subject of legal enactment. But it is to be regretted that such legislation and its enforcement have never given the protection that should have been afforded. Our legislators, in general, are not adequately informed as to the utility of birds, else they would hardly be willing to leave any of them beyond the pale of legal protection. Many birds seemingly useless, or even destructive, will be found, on closer observation, to be among our greatest benefactors. Natural history, in general, and Ornithology, in particular, are interesting and important branches of education.

Classification of Birds.

The Systematic Classification of birds is attended with peculiar difficulties, owing to their many resemblances, combined with manifold variations. Numerous systems of classification have been devised, especially since the beginning of the present century; but none of these are entirely satisfactory, all of them being considered, by Ornithologists in general, as more or less unnatural and artificial. The system of Linnæus has been long abandoned, although its author deserves the same credit for his orderly distributions in Ornithology as in other departments of Zoology.

In the classification herein attempted, the suggestions of the best modern Ornithologists have been adopted. There is doubtless room enough for many improvements; and improvements will certainly be made, as the further progress of the Science of Birds shall lead to agreement, on doubtful or disputed points, among its votaries.

FIRST CLASS.

CRACKERS—(Enncleatores)—Three Orders.

FIRST ORDER.

Parrots—(Psittacini)—Six Groups.

First Group—Parrots Proper—(Psittaci)—Seven Families.

1. Gray Parrots—Psittacus.
2. Green Parrots—Chrysotis.
3. Short-tail Parrots—Pionus.
4. Fan Parrots—Deroptyus.
5. Dwarf Parrots—Agapornis.
6. Sparrow Parrots—Psittacula.
7. Siskin Parrots—Nasiterna.

Second Group—Loris—(Lorii)—Four Families.

1. Loris—Lorius.
2. Lorikets—Psittacuteles.
3. Purple Loris—Coryphilus.
4. Long-tailed Loris—Pyrrhodes.

Third Group—Cockatoos—(Physolophi)—Seven Families.

1. Cockatoos—Cacatua.
2. Helmet Cockatoos—Calocephalus.
3. Fan Cockatoos—Licmetis.
4. Ground Cockatoos—Nestor.
5. Hair Cockatoos—Dasyptilus.
6. Proboscis Cockatoos—Microglossum.
7. Raven Cockatoos—Calyptorhynchus.

Fourth Group—Night Parrots—(Strigopes)—One Family.

1. Night Parrots—Strigops.

Fifth Group—Araras—(Ara)—Four Families.

1. Araras—Ara.
2. Blue Araras—Anodorhynchus.
3. Wedge-tail Araras—Conurus.
4. Nose Araras—Euicognathus.

Sixth Group—Sittiches (Sparrow or Finch Parrots)—Palaeornides)—Seven Families.

1. Noble Sittiches—Palaeornis.
2. Superb Sittiches—Polytelis.
3. Grass Sittiches—Platycercus.
4. Speckled Sittiches—Psephotus.
5. Ornamental Sittiches—Melopsittacus.
6. Painted Sittiches—Nymphicus.
7. Ground Sittiches—Pezoporus.

SECOND ORDER.

SPARROW BIRDS—(*Passeres*)—Thirteen Groups.

First Group—Crossbills—(*Loxia*)—Two Families.

1. Crossbills—*Loxia*.
2. Siskin Finches—*Psittirostra*.

Second Group—Bullfinches—(*Pyrrhula*)—Eight Families.

1. Parrot Bullfinches—*Paradoxornis*.
2. Pine Bullfinches—*Pinicola*.
3. Redbreast Bullfinches—*Erythrothorax*.
4. Long-tail Bullfinches—*Uragus*.
5. Desert Bullfinches—*Bucanetes*.
6. Red Bullfinches—*Pyrrhula*.
7. Garden Bullfinches—*Serinus*.
8. Tree Bullfinches—*Dryospiza*.

Third Group—Finches—(*Fringilla*)—Eight Families.

1. Noble Finches—*Fringilla*.
2. Alps Finches—*Montifringilla*.
3. Winter Finches—*Niphœa*.
4. Hemp Finches—*Cannabina*.
5. Birch Siskins—*Linaria*.
6. Siskins—*Spinus*.
7. Thistle Siskins—*Carduelis*.
8. Goldfinches—*Astragalinus*.

Fourth Group—Sparrows—(*Passeres*)—Four Families.

1. Sparrows—*Passer*.
2. Sparrowlets—*Pyrgitopsis*.
3. Gold Sparrows—*Chrysospiza*.
4. Mountain Sparrows—*Petronia*.

Fifth Group—Grosbeaks—(*Coccothraustes*)—Four Families.

1. Green Finches—*Chloris*.
2. Grossbeaks—*Coccothraustes*.
3. Evening Grosbeaks—*Hesperiphona*.
4. Nut Grosbeaks—*Geospiza*.

Sixth Group—Parrot Finches—(*Pityli*)—Nine Families.

1. Kernel Crackers—*Coccoborus*.
2. Crested Kernel Crackers—*Cardinalis*.
3. Gray Finches—*Parvaria*.
4. Bullfinch Finches—*Sporophila*.
5. Ruider Finches—*Catamblyrhinchus*.
6. Parrot Finches—*Pitylus*.
7. Mask Finches—*Coriothraustes*.
8. Habias—*Saltator*.
9. Plant-mowers—*Phytotoma*.

Seventh Group—Color Finches—(*Tanagra*)—Six Families.

1. Tanagers—*Tanagra*.
2. Fire Tanagers—*Pyranga*.
3. Calliots—*Calliste*.
4. Callions-billed Tanagers—*Rhamphocelus*.
5. Strangling Tanagers—*Lavia*.
6. Organists—*Euphone*.

Eighth Group—Pomp Finches—(*Amadina*)—Twelve Families.

1. Collar Finches—*Amadina*.
2. Cap Finches—*Spermestes*.
3. Cracker Finches—*Pyrenestes*.
4. Bullrush Finches—*Donacola*.
5. Grass Finches—*Poëphila*.
6. Rice Finches—*Padda*.
7. Gold Finches—*Pytelia*.
8. Drop Finches—*Lagonosticta*.
9. Variegated Finches—*Emblema*.
10. Lustre Finches—*Hypochera*.
11. Butterfly Finches—*Mariposa*.
12. Astrilds—*Astrilda*.

Ninth Group—Weaver-Birds—(*Ploceï*)—Seven Families.

1. Social Weavers—*Philetærus*.
2. Gold Weavers—*Ploceus*.
3. Bunting Weavers—*Nelicurvius*.
4. Bloodbill Weavers—*Quelea*.
5. Mourning Weavers—*Taha*.
6. Fire Weavers—*Euplectes*.
7. Cattle Weavers—*Textor*.

Tenth Group—Widow Birds—(*Vidua*)—Five Families.

1. Mourning Widows—*Colinspasser*.
2. Train Widows—*Chera*.
3. Cocktail Widows—*Steganura*.
4. Widows—*Vidua*.
5. Hairtail Widows—*Tetraxera*.

Eleventh Group—Bunting Finches—(*Passerella*)—Four Families.

1. Morning Finches—*Zonotrichia*.
2. Bunting Finches—*Spizella*.
3. Steps Finches—*Passerculus*.
4. Short Finches—*Ammodramus*.

Twelfth Group—Buntings—(*Emberiza*)—Seven Families.

1. Cardinal Buntings—*Gubernatrix*.
2. Gray Buntings—*Miliaria*.
3. Buntings—*Emberiza*.
4. Pomp Buntings—*Euspiza*.
5. Bullrush Buntings—*Cynchramus*.
6. Spur Buntings—*Centrophanes*.
7. Winter Buntings—*Plectrophanes*.

Thirteenth Group—Larks—(*Alauda*)—Ten Families.

1. Kalander Larks—*Melanocorïpha*.
2. Steppes Larks—*Saxilauda*.
3. Sand Larks—*Ammomanes*.
4. Bunting Larks—*Pyrrhalauda*.
5. Shore Larks—*Phileremos*.
6. Tuft Larks—*Galerita*.
7. Wood Larks—*Corys*.
8. Larks—*Alauda*.
9. Spur Larks—*Macronyx*.
10. Runner Larks—*Alamon*.

THIRD ORDER.

RAVEN BIRDS—(*Coracirostres*)—Eight Groups.

First Group—Starlings—(*Icteri*)—Seven Families.

1. Rice Eaters—*Dolichonix*.
2. Swamp Trupials—*Agelaius*.
3. Cow Birds—*Molothrus*.
4. Yellow Birds—*Icterus*.
5. Hang-Nests—*Hyphantes*.
6. Blackbirds—*Cassicus*.
7. Boat-tails—*Quiscalus*.

Second Group—Starlings Proper—(*Sturni*)—Five Families.

1. Starlings—*Sturnus*.
2. Starlingouzel—*Pastor*.
3. Shepherd Starlings—*Acridotheres*.
4. Grakles—*Gracula*.
5. Maggot Choppers—*Buphaga*.

Third Group—Lustre Thrushes — (*Lamprotornithes*) — Six Families.

1. Lustre Starlings—*Lamprocolius*.
2. Starling Lustre Thrushes—*Notauges*.
3. Scale Lustre Starling—*Pholidauges*.
4. Lustre Stars—*Lamprotornis*.
5. Rock Lustre Birds—*Pilorhinus*.
6. Mountain Lustre Starlings—*Amydrus*.

Fourth Group—Oriole and Orioles—Four Families.

1. Saw Birds—*Prionoderus*.
2. Collar Birds—*Sphecotheres*.
3. Oriols—*Oriolus*.
4. Silk Oriols—*Sericulus*.

Fifth Group—Birds of Paradise—(*Paradisea*)—Seven Families.

1. Birds of Paradise—*Paradisea*.
2. Cinchoobirds—*Cinnamura*.
3. Collared Birds of Paradise—*Lophorina*.
4. Ornamented Birds of Paradise—*Parotia*.
5. Pomp Hoopoes—*Seleucides*.
6. Ornamented Hoopoes—*Epimachus*.
7. Paradise Pies—*Astrapia*.

Sixth Group—Ravens — (*Coraces*)—Thirteen Families.

1. Rock Crows—*Fregilus*.
2. Rock Jackdaws—*Pyrrhocorax*.
3. Noble Ravens—*Corax*.
4. Vulture Ravens—*Corvultur*.
5. Ornamented Ravens—*Picicorvus*.
6. Crows—*Corvus*.
7. Field Crows—*Frugilegus*.
8. Tower Crows—*Monedula*.
9. Pretty Crows—*Anomalocorax*.
10. Pine Crows—*Nucifraga*.
11. Piping Crows—*Streperidae*.
12. Piping Pies—*Strepera*.
13. Bald Crows—*Phoenotis*.

Seventh Group—Jays—(*Garruli*)—Eleven Families.

1. Garden Jays—*Pica*.
2. Pie Jays—*Cyanopica*.
3. Tree Jays—*Cyanocorax*.
4. Strangling Jays—*Cyanocitta*.
5. Jays—*Garrulus*.
6. Lichen Jays—*Perisoreus*.
7. Tail Jays—*Dendrocitta*.
8. Lobe Jays—*Crypsirhina*.
9. Stump Jays—*Temnurus*.
10. Kittas—*Urocissa*.
11. Feather Bills—*Cissa*.

Eighth Group—Plantain Eaters—(*Amphibolae*)—Five Families.

1. Plantain Eaters—*Musophaga*.
2. Helmet Birds—*Corythaix*.
3. Turakos—*Corythaeola*.
4. Bustle Birds—*Schizorhis*.
5. Mice Birds—*Colius*.

SECOND CLASS

CATCHERS—(*Captatores*)—Three Orders.

FOURTH ORDER.

Birds of Prey—(*Raptores*)—Thirteen Groups.

First Group—Falcons—(*Falcones*)—Eight Families.

1. Hunting Falcons—*Hierofalco*.

2. Peregrine Falcons—*Falco*.
3. Lark Falcons—*Hypotriorchis*.
4. Bush Falcons—*Hieracidea*.
5. Jogging Falcons—*Tinnunculus*.
6. Evening Falcons—*Erithropus*.
7. Throat Falcons—*Rhynchodon*.
8. Dwarf Falcons—*Harpa*.

Second Group—Hawks—(*Accipitres*)—Six Families.

1. Noble Hawks—*Herpetotheres*.
2. Tooth Hawks—*Harpagus*.
3. Sparrow Hawks—*Nisus*.
4. Hawks—*Astur*.
5. Song Hawks—*Melierax*.
6. Snake Hawks—*Polyboroides*.

Third Group—Eagles—(*Aquilae*)—Ten Families.

1. Eagles proper—*Aquila*.
2. Buzzard Falcon Eagles—*Circaetos*.
3. Hawk Eagles—*Pandion*.
4. Hooded Eagles—*Spizaetos*.
5. Tuft Eagles—*Lophaetos*.
6. Strangling Eagles—*Pternura*.
7. Sparrow Eagles—*Morphnus*.
8. Rings Eagles—*Harpya*.
9. Sea Eagles—*Haliaetos*.
10. Fishing Eagles—*Pandion*.

Fourth Group—Kites—(*Circi*)—Twelve Families.

1. Eagle Kites—*Helotarsus*.
2. Swimmers—*Elanus*.
3. Hover Kites—*Ictinia*.
4. Buzzard Kites—*Cymindis*.
5. Hood Kites—*Baza*.
6. Water Kites—*Hydroictinia*.
7. Kites—*Milvus*.
8. Swallow Kites—*Nauclerus*.
9. Dwarf Swallow Kites—*Chelidopteryx*.
10. Field Kites—*Strigiceps*.
11. Swamp Kites—*Circus*.
12. Speckled Kites—*Spilocircus*.

Fifth Group—Buzzards—(*Buteones*)—Eight Families

1. Eagle Buzzards—*Circaetos*.
2. Speckled Buzzards—*Spilornis*.
3. Honey Buzzards—*Pernis*.
4. Rough Legged Buzzards—*Archibuteo*.
5. Buzzards—*Buteo*.
6. Stepps Buzzards—*Poliornis*.
7. Hook Buzzards—*Rostrhamus*.
8. Heel Buzzards—*Hypomorphnus*.

Sixth Group—Vulture Falcons—(*Polybori*)—Three Families.

1. Vulture Buzzards—*Milvago*.
2. Vulture Falcons—*Polyborus*.
3. Crying Buzzards—*Ibycter*.

Seventh Group—Crane Vultures—(*Gypogerani*)—One Family.
1. Crane Vulture or Secretary—*Gypogeranus serpentarius*.

Eighth Group—Vulture Eagles—(*Gypaëti*) One Family.
1. Lämmergeier, Barbed Vulture—*Gypaëtos barbatus*.

Ninth Group—Vultures—(*Vultures*)—Four Families.

1. Comb Vultures—*Sarcorhamphus*.
2. Goose Vultures—*Gyps*.
3. Tuft Vultures—*Vultur*.
4. Ear Vultures—*Otogyps*.

Pl. X.

1 ŒDICNEMUS crepitans   2 ESACUS magnirostris

CURSORINÆ.

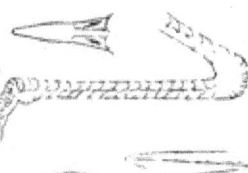

1 PLUVIANUS ægyptius   2 CURSORIUS gallicus   3 OREOPHILUS ruficollis

GLAREOLINÆ.

The third Subfamily,

GLAREOLINÆ, or PRATINCOLES,

have a short Bill, which is broad at the base, and laterally compressed to the tip; the Wings very long, with the first quill the longest; the Legs moderate, with the tip of the tibia naked; the Toes three in front, and one posteriorly, which is elevated.

GLAREOLA Briss.*

Bill short, broad at the base, much compressed to the tip, with the culmen depressed at the base, elevated and arched to the tip, the lateral margins curved; the nostrils basal, lateral, and oblique. *Wings* lengthened, pointed, extending beyond the end of the tail, with the first quill longest. *Tail* moderate, and more or less forked. *Legs* moderate and slender, with the tarsi scutellated, and the middle toe and claw lengthened; the outer toe longer than the inner, and united at the base to the middle one; the hind toe very short, elevated, but touching the ground; and the claws rather long, nearly straight, that of the middle toe slightly pectinated on one side.

These few species inhabit the temperate and warmer parts of the Old World. They frequent the borders of rivers, lakes, and marshes, both in the plains and on the mountains. Their food consists chiefly of worms, flies, orthopterous and aquatic insects, which they take on the wing like the swallows, and on the ground, where they can run very quickly. They form a slight nest on the surface of the ground, among the rushes and thick herbage in the marshes, wherein they deposit several eggs.

1. G. *pratincola* (Linn.) Pall. Pl. enl. 882. — Glareola austriaca Gmel. Leach, Linn. Tr. xiii. pl. 12.; Glareola nævia Gmel.; Glareola senegalensis Gmel.; Glareola torquata Meyer.

2. G. *Nordmanni* Fisch. — Glareola pratincola Pall.

3. G. *limbata* Rüpp.

4. G. *orientalis* Leach, Linn. Tr. xiii. pl. 13.

5. G. *isabella* Vieill. Gal. des Ois. t. 263. — Glareola grallaria Temm.; Glareola australis Leach, Linn. Tr. xiii. pl. 14.

6. G. *lactea* Temm. Man. ii. 503., Pl. enl. 399.

7. G. *cinerea* Fras. Proc. Z. S. 1843. 26.

* Brisson established this genus in (*Ornithologie*) 1760, and in 1777 Scopoli proposed *Trachelia*.

GLAREOLA
cinerea. Fraser

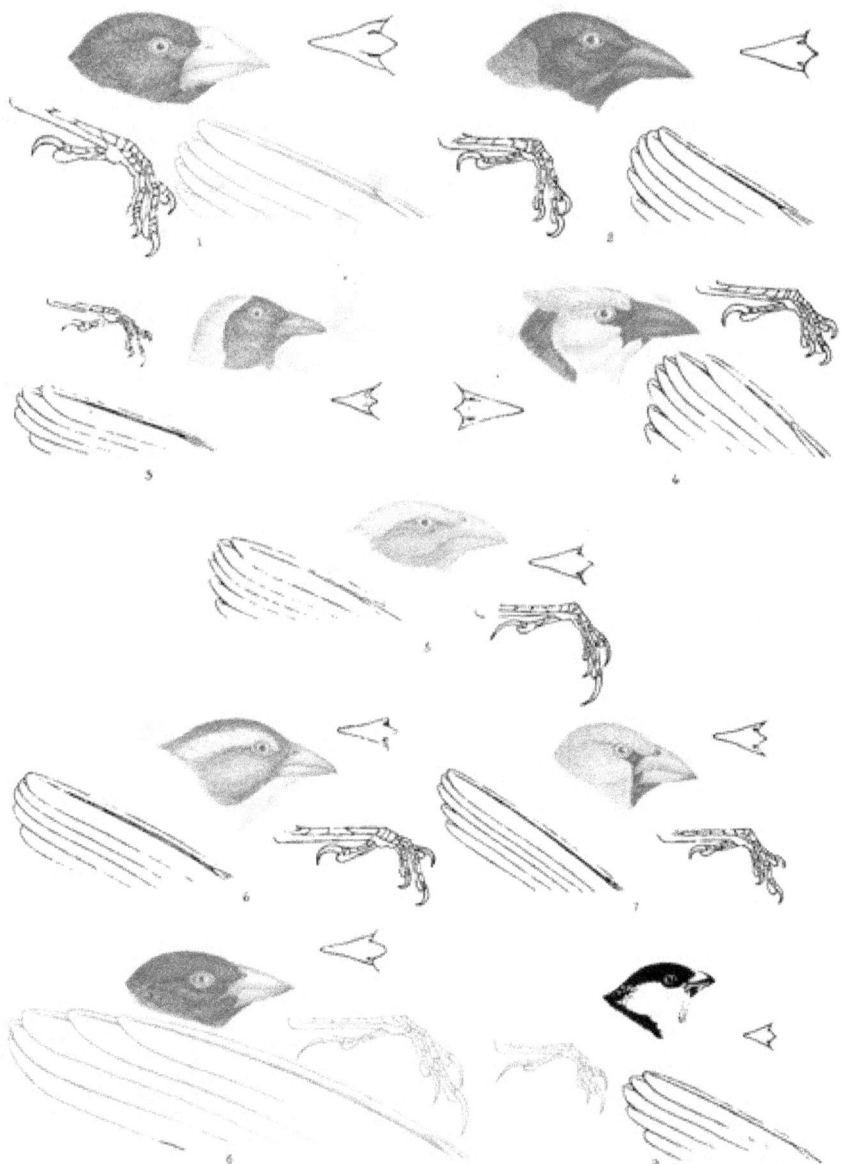

PLOCEINÆ.

1 TEXTOR alecto   2 HYPHANTORNIS textor   3 NIGRITA canicapilla   4 SYCOBIUS cristatus
5 PLOCEUS philippinus   6 PLOCEPASSER mahali   7 PHILETÆRUS socius   8 CHERA progne
9 VIDUA principalis

Order VII. GRALLÆ *Linn.* *

comprehends a large series of birds that have the lower portion of their Tibiæ, or Thighs, naked, and the Tarsi lengthened, rounded, and slender.

The first Family,

CHARADRIADÆ, or Plovers,

have the Bill short, with the basal portion of the culmen rather depressed and weak, and the apical part strong and swollen; the Nostrils placed in a deep longitudinal groove of various length; the Tarsi lengthened; the hind Toe totally wanting, or small and elevated.

The first Subfamily,

ŒDICNEMINÆ, or Thick-knees,

have the Bill as long as, or longer than, the head, with the culmen slightly depressed at the base and swollen at the tip, and the gonys more or less angulated; the Tarsi lengthened, with three rather short Toes in front.

ŒDICNEMUS *Temm.* †

Bill rather longer than the head, the culmen straight, with the apical half arched and curved to the tip, the sides compressed, and the gonys nearly half the length of the bill, angulated, and advancing upwards to the tip; the nostrils in a subtriangular membranous groove, with the aperture longitudinal and anterior. *Wings* of moderate length, pointed; with the first quill shorter than the second, which is the longest, and the tertials the length of the quills. *Tail* moderate and wedge-shaped. *Tarsi* lengthened, three or four times the length of the middle toe, and covered with hexagonal scales. *Toes* short, the inner shorter than the outer, and both united to the middle one by a membrane at their base, especially the outer; the claws short and slightly curved.

---

* Or the *Gradiatores* of Illiger.
† This genus was established by M. Temminck in (*Manuel d'Ornithologie*, 1st edit. p. 321.) 1815; and the *Fedoa* of Leach, proposed in 1816, is coequal.

They are migratory birds, inhabiting all parts of the world except North America, seeking the more temperate regions to rear their young, and the warmer latitudes to pass the winter. These periodical flights are performed in flocks during the night, with great swiftness. It is in uncultivated open moorlands that these birds are generally found. Their food is sought for during the evening or at night; it consists of small quadrupeds, reptiles, and especially worms and insects. During the day they sit closely squatted behind a stone, or any other object sufficiently large to hide them; but, if disturbed, they fly to a short distance, and then run off to hide with great rapidity. Each female deposits two eggs on the surface of the bare ground. The young are capable of following the parent as soon as they are excluded from the egg.

1. Œd. crepitans Temm. Pl. col. 919. — Charadrius Œdicnemus Linn.; Œd. europeus Vieill.; Œd. griseus Koch.

2. Œd. senegalensis Swains. Birds of W. Afr. ii. 228. — Œd. affinis? Rüpp. Mus. Senck. 1834. 210.

3. Œd. maculosus Temm. Pl. col. 292. — Œd. capensis Licht.

4. Œd. bistriatus (Wagl.) Isis, 1829. 648.—Œd. vocifer L'Herm. Mag. de Zool. 1837. pl. 84.; Œd. americanus Swains.

5. Œd. grallarius (Lath.) Lambert's Icon. ined. iii. t. 15.—Œd. longipes Vieill. Pl. col. 386.; Charadrius frenatus Lath. Lambert's Icon. ined. iii. t. 40.; ? Charadrius magnirostris Lath. Lambert's Icon. ined. ii. t. 19.

6. Œd. giganteus Licht. Isis, 1829. 647.

Esacus Less. †

Bill much longer than the head, strong, the culmen more or less straight, with the base cultrated, and the tip gradually or suddenly hooked; the base broad, and the sides gradually compressed to the tip; the lateral margins more or less curving upwards to the tip, and angulated at the base; the lower mandible strong, with the gonys half its length, angulated, and advancing upwards to the tip; the nostrils placed in a membranous groove, rather less than half the length of the bill, with the aperture longitudinal, anterior, and near the margin.

They inhabit the wide sandy banks of the larger rivers of India during the winter, and, as the summer advances, migrate to the northern parts of India. Their food consists of crabs and other hard shellfish. They are also found in the Indian Archipelago and Australia.

1. Es. recurvirostre (Cuv.) Less. — Curvacus griseus Hodgs.; Œdicnemus recurvirostris Swains.

2. Es. magnirostris (Geoff.) Temm. Pl. col. 387.

* The type of Illiger's genus Burhinus, which was established in 1811 on Latham's short description, taken from the badly executed drawing referred to above.

† This is coequal with Esacus, of Mr. Hodgson, published in the Journ. As. Soc. Beng. 1836, p. 776. In 1841 he changed it to Ponoceps. M. Lesson's name was published in 1831, in his Traité d'Ornithologie, p. 547.

May, 1844.

Pl. XI.

Pl. XII.

tree, like a little rest, and, having quickly picked up some food, go to sleep. Next morning after a slight breakfast, the journey is resumed. The movement of these birds is pleasant and easy : their flight straight, and their walk on the ground quiet. They manifest great skill in climbing branches ; in this respect almost surpassing the Titmouse.

The Orchard Oriole. (Oriolus—Icterus—Mutatus.)

Fig. 3.

This bird chiefly frequents orchards, whence the name. It is gay and frolicksome, and seems to be always in great haste, hopping among the branches or upon the ground, and flying in the air. Its notes are short but lively, and uttered with such rapidity that it is difficult to follow them distinctly. Sometimes it utters only a single note, which is very agreeable. Its food generally consists of insects and their larvæ. Of the insects that infest fruit trees, they destroy great quantities, and are therefore benefactors to farmers and fruit-growers.

The Orchard Oriole builds his nest similar to that of the Baltimore. For material it uses a long fibrous grass, and generally hangs the nest on the outward branch of an apple tree. The nest is semi-globular in shape, about three inches deep and two wide ; the inside is lined with wood or a down from the seeds of the platanus occidentalis, or buttonwood tree. The eggs are commonly four in number, having a pale bluish tint, with a few small specks of brown and dots of purple. The female sits fourteen days ; the young remain from two to three weeks in the nest, which they leave about the middle of June. The upper portion of the female is colored with a yellowish olive, inclining to a brownish tint on the back ; the wings are dusky brown, and the lesser wing coverts tipped with yellowish white ; the tail is rounded, the two exterior feathers three-quarters of an inch shorter than the middle ones ; the lower parts of the body are yellow. The plumage of the male nearly corresponds with that of the female.

The Indigo Blue Bird. (Cyanospiza Cyanea.)

Fig. 4.

This beautiful little bird inhabits, it seems, all parts of the North American continent from Mexico to Nova Scotia, and from the sea-coast west, beyond the Appalachian and Cherokee Mountains. It is chiefly seen in gardens, fields of clover, on the borders of woods, and on roadsides, where it is often observed perched on fences. It is very neat and agile, and a good singer. Mounting to the highest top of a tree it sometimes chants for half an hour at a time. Its song consists of short notes often repeated ; the first ones are loud and rapidly uttered each other ; then they are gradually dropped until they are hardly audible, the little singer appearing to be quite exhausted ; but after a pause of about half a minute, he begins again as fresh, lively, and loud as at first. The song is heard during the months of May, June, July, and August. When frightened it utters a single chirp, sounding almost like two pebbles struck together. The color of its plumage is changeable, depending on the light in which it is seen. Instead of indigo blue, it sometimes appears in a verdigris dress ; at other times the dress appears green, and at others blue. Its head is of a deep blue, and its color is not changeable like that of the rest of the body. Its nest is usually both in rank grass, grain, or clover, and is generally suspended between two twigs, one passing on each side ; it is composed of flax or other fibrous material, with an inside lining of fine dry grass. The eggs, numbering five, are light blue, with a purplish blotch on the larger end. Insects and a variety of seeds constitute its principal food. The female is of a light flaxen color ; her wings are of a dusky black, and the cheeks, breast, and the lower portions

of her body are clay-colored, streaked with a darker color under the wings, tinged so as to be bluish in several places. Toward fall, after moulting, the male appears almost in the same colors as the female. The Indigo Blue Bird is frequently kept in cages ; and those taken in trap-cages soon become reconciled to their captivity, but never sing so well nor so loud as those reared by hand from the nest. They are fed with different kinds of seeds, such as rape, turnip, hemp, and canary seed.

In Europe they are invariably found in every collection of birds.

The Hooded Fly-catcher. (Muscicapa—Setophaga Mitrata.)

Fig. 5.

This bird is chiefly found in the southern parts of North America, abounding in the Gulf States. It is a lively bird, and has in a good degree the manners of a true Fly-catcher, while in some respects it resembles the Warbles. It is in an almost constant chase after insects, its principal food, uttering now and then a very lively " tweet, tweet, twitchee." In the Northern States it is rather scarce, and when met with there it is shy and timid, like a stranger far from home.

It spends the winter in Mexico and the West India islands. The nest of the Hooded Fly-catcher is very neatly and compactly built in the fork of a small bush ; it is on the outside composed of flax and other fibers, and moss, or pieces of broken hemp ; the inside is nicely lined with hair and feathers. The eggs are five in number, grayish white, with reddish spots on the larger end. In the United States it is a bird of passage. The female nearly resembles the male, except that the yellow of her throat and breast has a slight blackish tint ; the black does not reach so far down on the upper part of the neck as in the male, and it is also of a less deep color.

───────

PLATE XIII.

Townsend's Cormorant. (Phalacrocorax Townsendii.)

Fig. 1.

Cormorants are generally found in all parts of both hemispheres ; in middle Asia, and, in winter, in great numbers in Africa. They are most numerous in rivers bordered by large forests. Thousands congregate on the Columbia river. The bird from which the drawing is made, was presented to me by Dr. W. T. Shepard, who shot it in the "Reservoir," in Licking county, Ohio. It proved, on dissection, to be a female.

Cormorants are creatures in winter in all the southern seas—in Greece, in China, and India. Wherever water and fish are to be met with, Cormorants are seen. These birds manifest many peculiarities. They are gregarious, usually congregating in flocks, and sometimes in considerable numbers. They are seldom seen singly or in pairs. Almost all the different kinds of Cormorants are often collected in the same flock.

During the morning hours, Cormorants are busy in fishing. The afternoon is generally devoted to repose. Toward evening another fishing excursion is made, and after this they retire to sleep. For this purpose they select, in the interior of the country, high trees on islands, or those standing in lakes or rivers. Such trees also serve them for breeding-stations. On the coast or on the ocean, they choose a rocky island, affording a wide range of vision, and also a harbor, from whose every side they can easily take flight and return. Such islands can be seen and recognized from a distance, as they are literally covered with the white excrements of these birds. Ship-loads of guano could be collected on these islands, if it could only be dried by the tropical sun of Peru. Such a sight in mid-ocean never fails to attract the attention of the mariner or the traveler; but the island is, of course, most attractive when it is occupied by Cormorants. There they sit arranged in rows or lines, on the rocks, in the most picturesque positions, and all facing the sea. Rarely can one be seen sitting apart from the rest. They usually wear a stiff, statue-like appearance; but sometimes each bird is seen to move some part of the body, either the neck, wings, or tail. The object of these movements doubtless is to dry their feathers. After ten or fifteen minutes, they become quiet, merely basking in the sun. On such occasions, each Cormorant seems to have a particular place which he always occupies.

Cormorants walk with extreme difficulty. Some observers have said that these birds can only walk when they support themselves by their tails. This supposition has evidently arisen from the fact that the tail portion of the Cormorant's body is stiff, like that of the Woodpecker. Cormorants, when hanging by their short, round claws at the entrances to crevices or hollows in rocks, support themselves by their tails as Woodpeckers do. The walk of Cormorants is a mere waddling, and yet they make more rapid progress than an observer would at first sight suppose. They are not made for locomotion on land; but in swimming and diving they are experts. When a boat approaches their resting-place, they stretch out their necks, take a few irregular steps, and turn as if for a general flight; but only a few take to flying, bravely flapping their wings for a short time. These maneuvers are followed by a regular sail in the air; while others fly round in circles, rising higher and higher like the Hawk or Kite. The majority, however, do not take to the wing at all, but let themselves down into the water, head foremost, like frogs, diving and rising at a great distance off. Then, looking for a moment at the boat with their green eyes, they dive and rise again, and so keep doing till they reach a place of safety.

There is probably no bird that can surpass the Cormorant in diving and swimming under water. Frequent trials have been made to get ahead of them with a light boat or canoe; but the practiced oarsman, though exerting himself to the utmost, could make only half the distance on the surface that the Cormorants made in the same time under water. They dive to great depths, and remain a long time under water; then coming up to the surface, they hastily draw in a fresh supply of air and dive again. When pursuing their prey in the water, they stretch themselves out and swim with sturdy strokes, pushing themselves through the water with an arrow-like velocity.

It may be reasonably inferred from the penetrating green eyes of Cormorants that their sense of vision is well developed. Their hearing is also acute, and they do not lack the sense of feeling. But they are too voracious to possess much discrimination in the sense of taste. It is true they feed on one kind of fish more than on any other; but this preference is probably not so much due to their taste, as to the fact that such fish are more easily caught than others. The fish alluded to is the so-called alewife, a kind of herring, found in great numbers, swimming near the surface. Cormorants are shy and distrustful. Toward other birds, with whom they come in contact, their behavior is that of tricksters and rascals.

The Chinese train Cormorants for fishing. The young intended for this use are hatched by domesticated hens. The following is the mode of fishing with Cormorants: The fisherman employs a raft from fifteen to twenty feet in length, and from two and a half to three feet in width, made of bamboo, and furnished with an oar and rudder. Arriving on the fishing ground, he drives the Cormorants from the raft into the water, and they all dive at once. As soon as a Cormorant has caught a fish, rising with it to the surface, he swims toward the raft, merely with the intention of swallowing the fish. He is prevented by a brass ring or string around his neck from accomplishing this feat. The fisherman hurries toward the bird, throws a net over him, drags him to the raft, and secures the fish. He then sends the Cormorant back into the water for more booty.

In the interior of a country, Cormorants in a very short time destroy all the fish in the lakes and rivers. Their voracity exceeds comprehension. A single Cormorant devours daily from sixteen to twenty good-sized herring. They catch, it is said, young aquatic birds, Ducks, Coots, Rails, etc. The writer has found in a Cormorant's stomach the remains of a young Gallinula.

Cormorants prefer trees for nest-building, but also make use of hollows in rocks. Their nests are formed of a few dry rushes, fibrous roots, etc. Crows and Herons are often expelled from their nests by Cormorants, who appropriate the nests to their own use. Toward the close of April, the female Cormorant lays three or four bluish green eggs, of an oblong shape, and small in proportion to the size of the bird. The male and female sit alternately on the eggs, and usually hatch them out in about twenty-eight days. They also take turns in feeding the young. These grow rapidly, and are well taken care of by their parents, who, however, do not try to defend them, at least not against man. On arriving at the nest from a fishing excursion, the parent birds empty their crops and stomachs, which sometimes contain several dozen small fishes. Many of these fall over the border of the nest to the ground; but the Cormorants never take the trouble to pick them up. Toward the middle of June the young are able to fly, and the old birds begin raising a second brood. The flesh of Cormorants is not generally considered fit for food; but Laplanders and other northern people pronounce it delicious.

The Double-crested Cormorant. (Phalacrocorax Dilophus.)

Fig. 2.

This bird is represented on the plate in its summer plumage, having two elongated tufts of feathers behind each eye. It inhabits all parts of this country from Maryland to Labrador, but in no way differs from other Cormorants. The specimen that served for the drawing, was shot in the "Licking Reservoir," heretofore referred to, among a flock of the common Cormorants (Phalacrocorax Carbo).

PLATE XIV.

The Great Northern Diver Loon. (Colymbus Glacialis.)

Fig. 1.

The great Northern Diver, Loon, or Stater, as this bird is called in northern Europe, is a regular sea-bird, living on the coast, but frequenting large fresh-water lakes and ponds in the interior for the purpose of breeding. These birds, on their migration southward, late in the fall, and on their return northward, in April or May, visit our rivers and mill-ponds. They are very shy, wary, and difficult to kill, eluding the sportsman by their astonishing dexterity in diving and swimming under water, even against the current. They can remain a good while beneath the surface, often six or eight minutes at a time, and swim long distances with incredible rapidity, and without any apparent exertion. They sometimes lie flat on the surface of the water, or sink themselves in it, so that only a small portion of their backs and their heads and necks can be seen. They sometimes swim in a slow, quiet way. Their diving is accomplished without making any noise, or any commotion in the water, by stretching themselves up, bending the neck in a curve forward, and then plunging down. Under water they stretch out to their full length, press wings and feathers close to the body, and, moving their feet only, shoot onward like an arrow through the water. Sometimes they swim in one direction, and then in another; sometimes just beneath the surface, and then at a depth of several fathoms. They swim or race with fish, their usual food, and catch them while swimming. From the very first day of their lives, they swim and dive, and seem to feel safer in water than when flying high in the air.

These birds are quite helpless on the surface of the ground, which they avoid as much as possible. They can not walk as other birds do, or even hardly stand upright. They crawl along instead of walking, supporting themselves by their bills and using their wings to aid a forward movement. Their flight is much better than one would suppose it could be, with their heavy bodies and small wings. To get fairly on the wing, they make a long preliminary movement; but as soon as they have gained a certain height, they speed quickly forward, although compelled to flap their short wings in rapid succession. Loons are distinguished from all other sea-birds by their loud and sonorous voice. Many ornithologists speak of the voice as harsh and disagreeable; but the writer can not avoid confessing to a partiality for the loud morning call of the Loon. Its voice, especially at night, resembles a long drawn out "Auweek! Auweek?" So penetrating is it as sometimes to produce an echo in the surrounding rocks or mountains, sounding like the cry of a man in imminent peril of life.

Loons are shy and cautious, trusting no one. Strange creatures they avoid as much as possible, and do not seem to care much even for their own kind. They are often found single, and, during the breeding season, in pairs, greatly attached to each other. It is seldom that two pairs are seen on the same pond, and more rarely still can even a single pair be seen on a pond occupied by other birds. During their migrations, or when in captivity, they always keep at a distance from other birds, and snap at them if they come near. When brought to bay, Loons fiercely defend themselves, inflicting ugly wounds with their strong, sharp bills.

They swallow small fish whole; but, as such as are of the size of the herring cause them trouble, larger ones are torn into small pieces and so devoured. It has been observed that captive Loons never pick up a dead fish; while freshly caught birds, placed in a large reservoir well stocked with fish, commence immediately to dive, chase, and catch and eat the fish. Fishermen on Lake Erie are in the habit of inclosing a small piece of water, there or four deep, with a kind of network reaching above the surface, for the purpose of keeping fish for market. Oftentimes, a Loon, attracted by the multitude of fish, alights in one of these inclosures, and is easily caught, as it can not again get on the wing, for want of a place from which to make its launch into the air.

These birds select for their breeding-places quiet fresh-water ponds or lakes, often preferring those situated at a considerable elevation above the level of the sea. During the breeding season, their food, sometimes roots, are oftener found than at other times. The nests are usually found on small islands, but in case there are no such islands, the birds build nests on the shore near the border of the rushes, constructing them of rushes and rank grass, carelessly put together. No attempt is made at concealment, and the female bird, sitting on the nest, can be seen from a great distance. She lays two eggs of an oblong shape, with a coarse-grained shell, and of an oil green color, sprinkled with dark gray and reddish brown specks and dots. Both the male and female sit alternately on the eggs, and mutually feed and take care of their offspring. The eggs are usually laid in the latter part of May, and the young are to be seen by the end of June. If food is lacking in the pond or lake where the nest is located, one of the parents takes care of the young while the other flies off to some point on a fishing excursion. As soon as the young birds are fledged, they leave the home of their infancy, and follow their parents to the larger lakes or the sea.

The flesh of the Loon is unfit for human food; it is rancid to the taste, and its odor is disgusting. The natives of Greenland use the skins of these birds for clothing, and the Indians about Hudson's Bay adorn their heads with curious of Loon feathers. Lewis and Clarke's exploring party saw, at the mouth of the Columbia river, robes made of Loon skins. While they wintered at Fort Clatsop, on that river, they observed great numbers of these birds.

The female is smaller than the male Loon. The bill is yellowish, and only the upper ridge and the top black, or of a blackish horn color; the crown, back, and part of the neck and the whole upper parts are pale brown; the plumage of a part of the back and scapulars is tipped with pale ash; the throat, lower side of the neck, and the whole underparts are white, but not so purely white as in the male, as these parts in the female have a dirty yellowish tinge. The quill feathers are dark brown. The female has neither the streaked bands on her neck, nor the white spots on her body.

The Tell-tale, Tattler, or Godwit. (*Totanus Melano Leucus.*)

Fig. 1.

This bird is well known to our gunners along the sea-coast and marshes. They stigmatize it with the name of Tell-tale, for its faithful vigilance in alarming the Ducks on the approach of the hunter, with its loud and shrill cry. This cry consists of four notes, uttered in rapid succession, and so loud and shrill as to alarm any Duck within hearing. But gunners, aware of this fact, look out in the first instance, for this bird, and often hush its warning voice forever, before it is aware of their stealthy approach.

This elegantly formed bird appears on our coasts about the beginning of April, breeds in the marshes, and leaves for the South in the middle of November. Not only do these birds build nests in salt-water marshes, but also in fresh-water swamps; sometimes on the dry ground, and even in an old stump. The nest is simply a hollow, made usually in a tussock of rank grass, inlaid with a few dry leaves of grass, a little moss, and with pine needles or leaves. The eggs, four in number, are proportionally large, pear-shaped, and of an oil green color, sprinkled with brownish gray specks and dots. The female bird hatches the eggs; but her mate is always at hand and on the watch. The young run about, following their parents, as soon as they are out of the shell, and conceal themselves, as all their kindred do, on the approach of danger, by lying flat on the ground, or in the grass or weeds. As soon as they are full-fledged, they look out for themselves, but remain with the old birds, flying at will from place to place, making longer and longer excursions, and at length, on some fine evening, setting out for a grand wandering tour.

In their winter-quarters, Tattlers associate with many other birds, but seldom form large flocks. It seems as if the company of strangers suited them better than that of their own kind. Their manner is pleasing; their walk elegant, quick, and striding, and their flight easy and rapid. They wade in deep water, and swim if necessary. They are generally seen, either searching for food or standing on the marsh, alternately raising and lowering the head, and, on the least approach of danger, uttering a shrill whistle, their warning cry, and then rising on the wing, generally accompanied by all the shore birds in the vicinity. Occasionally they rise to a great height, and their whistle can be distinctly heard, when the birds are beyond the reach of the eye. They become very fat in the fall, and their flesh is in high esteem for the table.

Nature seems to have intended this bird as a kind of guardian or sentinel for all other shore or aquatic birds. They feed on the shore, or in the bogs or marshes, with a feeling of perfect security, so long as the Tattler is at hand, and is silent; but the moment his whistle is heard, there is a general commotion, and directly not a bird is to be seen, the disappointed gunner, in his vexation, uttering between his teeth something the reverse of a prayer.

PLATE XV.

The Gray or Sea Eagle. (*Haliæetus Ossifragus Leucocephalus*)

This formidable Eagle lives in the same countries, on the same food, and frequents the same localities as the Bald or White-headed

Eagle, with which it often associates. In fact, the Sea Eagle so much resembles the Bald Eagle, in the form of the bill, in its size, in the shape of the legs and claws, differing from the latter only in color, that it seems at once to be the same bird, distinguished from the Bald Eagles previously observed simply by its age or stage of color. Another circumstance corroborating such an inference, is the variety of the colors of Sea **Eagles**; scarcely any two of them are found to be colored alike; the plumage of each being more or less shaded with light color or white. In some, the chin, breast, and tail coverts are of a deep brown, on others, these parts are much lighter, sometimes whitish, with the tail evidently changing in color, and merging into white.

In former times some of the best informed ornithologists asserted that Sea Eagles must be of a different kind from Bald Eagles, as, on examination of the nests of each, it was found that both the parent Sea Eagles are different in color from the parent Bald Eagles. But it takes the Bald Eagles full four **years** to perfect their plumage, though the younger ones begin to breed in the second year. These young ones passing for Sea Eagles, it is supposed that there are a great many more Sea Eagles than **Bald** or White-headed Eagles.

Almost everybody has heard or read stories of very young children having been seized and carried off by a Bald or Sea Eagle. But it is doubtful whether any of these tales relating tales would bear a very close or critical examination. While the writer was stopping at an inn in the Tyrol, the landlord courted the numerous attentions in great haste, and, opening a window, discharged his short rifle at a bird that was flying at so great a distance to be seen alarmed. He explained, by saying that he made it a point to kill, or at least to shoot at, every Lämmergeier that came to that spot, as one of them had carried off the child of his best friend. The name and residence of that friend having **been given**, he was visited, and the information imparted by him was, that a child had in reality been carried **off** by a Lämmergeier—but **one of his children**, as had been erroneously stated, **but** the child of **an innkeeper** residing some fifteen miles distant. On visiting the innkeeper, it was ascertained **that the story was** wholly without foundation in fact.

The Sea Eagle is a coward. The present writer once climbed to an Eagle's nest on a lofty yellow pine tree, standing near the bank of a small creek, in the southern part of the State of New York. During the progress of the climbing, the old Eagle flew about the tree, screaming and making a hissing sound, but keeping at a respectful distance from the climber. On reaching the nest, it was found to consist of a large pile of sticks, cornstalks, rushes, and some fibrous materials. The different layers showed that it had answered a similar purpose for several successive years. It contained two young Eagles that threw themselves at once upon their backs and showed fight when they saw him about looking at them, striking at him with their claws, making a piteous wailing with their beaks, opening them, and sometimes snuffing the air with a snap. Not once when their young were lifted out of the nest and examined, did the old Eagles venture to attack the intruder, though they sometimes came around him in a direct line, their open beaks with their loud screams all over, and seemingly in a terrible rage. But when with one or five years of the object of their stay, they suddenly turned off at a right angle, either to the right or left. After the young Eagles had been examined for a quarter of an hour, they were put back into the nest, and their course descended the tree, to the great relief of these affected and fussy parents.

PLATE XVI.

The Fish Hawk. (Pandion **Haliaetus.**)

The Fish Hawk bears also the name of Osprey, Fish Eagle, and **Fish Kite. Up** to the present time it has been regarded as belonging among the Eagles, from whom it differs in every respect. Its right position seems to be that of a connecting link between Eagles and Kites.

Fish Hawks are migratory birds, usually arriving on the North American lakes in the latter part of March, sometimes later, and departing during the closing days of September. They live exclusively on fish, and of course their haunts are where their food abounds. They build nests **on** high trees, constructed of stout sticks, rushes, moss, seaweed, etc. The female lays two, sometimes three, handsome, oblong eggs of a grayish white color, and speckled all over with light reddish dots.

Their long wings enable Fish Hawks to continue with ease a long time in the air. At the start on an excursion, they rise to a great height, and then letting themselves down gradually, they begin just above the level of the water (not inspection for fish. This inspection is not, however, carried upon while they in a wild keeping over the water. They come to the fishing-place by a slow motion circle, and so often, by continually looking about, whether any danger is to be apprehended. Alternately towering themselves and staying to a height of fifty or sixty feet, they sometimes poise themselves to take a better aim at a fish seen in the water, and then dart down with legs stretched forward in an oblique direction, disappearing for a short time in the water, and then reappearing on the surface, flapping their wings and shaking the water from their feathers. If unlucky, away they fly, to return and try their luck once again. Whether lucky or not, they usually leave the smaller pools after their first endeavor. Their peculiar mode of fishing necessitates the making of more a plunge to no purpose, but this does not at all discourage them, their motto always is, "Try again." They seldom make nest, except where, on their arrival at the North, they find the lakes and ponds still covered with ice.

When a Fish Hawk pounces upon a fish, he drives his claws into each, lows leaves back that they are not easily or very quickly withdrawn. Very often, miscalculating the size and weight of the fish, he endangers his own life, and sometimes loses it altogether by being drawn under the water by a heavy fish, and drowned. On the capture by this bird, there have been observed two talons on each side of the back. This is explained by the fact that no Fish Hawk can turn the outer toe either forward or backward, and that in seizing a fish, he turns this toe backward, so as to get a firmer hold. He does all his weight on each fish as he can conveniently carry, to bear upon them there at leisure and in safety, but he always lets he drags to the shore.

Fish Hawks are never known to attack quadrupeds or birds for the purpose of obtaining food. All species birds are as well acquainted with the Fish Hawk, that they are never alarmed at his approach. Grackles very often build their nests in the interstices between the sticks of the Fish Hawk's nests, and both kinds of birds live together in harmony. But other birds of prey, in the Wren or Bald Eagles, or Sea Eagles, torment the Fish Hawk. As soon as a Bald Eagle sees **the** Hawk with a fish, he closely attacks and compels the Fish Hawk to drop his **hard-earned booty**, which the robber Eagle seizes and appropriates to his own use.

Fish Hawks are greatly attached to their young, and defend them to their utmost against both men and birds of prey. One of the parents always remains near the nest, while the **other is out fish-ing.** It is remarkable that the tree **on** which the **nest** of a Fish Hawk is built, and where the young are reared, always withers and dies in a short time afterward. Whether this is owing to some poison imparted to the tree by the bird, or to the salt water constantly dripping from the heavy loads of the nest, or to some other cause, has not been satisfactorily learned.

On dissecting a Fish Hawk, there were found on the glands on the rump, which supply the bird with oil wherewith to lubricate its feathers in order to protect them from injury by being frequently wet. These glands were remarkably large, and contained a great quantity of white greasy matter as well as yellow oil. The gall was very small; but the intestines, with their numerous windings

D

Tenth Group—Raven Vultures—(*Catharta*)—Four Families.

1. Pharaons Vultures—*Percnopterus*.
2. Collar Vultures —*Neophron*.
3. Raven Vultures—*Cathartes*.
4. Crow Vultures—*Coragyps*.

Eleventh Group—Day Owls—(*Surnia*)—Five Families.

1. Falcon Owls—*Surnia*.
2. Hare Owls—*Nyctea*.
3. Rock Owls—*Athene*.
4. Burrow Owls—*Pholeoptynx*.
5. Sparrow Owls—*Microptynx*.

Twelfth Group—Eared Owls—(*Bubones*)—Four Families.

1. Great Horned Owls—*Bubo*.
2. Water Owls—*Kitupa*.
3. Eared Owls—*Otus*.
4. Dwarf Eared Owls—*Ephialtes*.

Thirteenth Group—Night Owls, Screech Owls—(*Striges*)—Three Families.

1. Tree Screech Owls—*Syrnium*.
2. Night Screech Owls—*Nyctale*.
3. Veiled Screech Owls, Barn Owls—*Strix*.

FIFTH ORDER

SPREADING BIRDS—(*Hiantes*)—Five Groups.

First Group—Swallows—(*Hirundines*)—Five Families.

1. Noble Swallows—*Cecropis*.
2. Rough-legged Swallows—*Chelidon*.
3. Gray Swallows—*Cotyle*.
4. Wood Swallows—*Atticora*.
5. Sailing Swallows—*Progne*.

Second Group—Sailors—(*Cypseli*)—Four Families.

1. Tree Sailor—*Dendrochelidon*.
2. Salanganes—*Collocalia*.
3. Prickle Sailors—*Acanthylis*.
4. Sailors—*Cypselus*.

Third Group—Night Swallows—(*Caprimulgi*)—Eight Families.

1. Day Shaders—*Podager*.
2. Twilight Swallows—*Chordeiles*.
3. Night Shaders—*Caprimulgus*.
4. Brittle Swallows—*Antrostomus*.
5. Train Swallows—*Scotornis*.
6. Water Swallows—*Hydropsalis*.
7. Deceit Swallows—*Macrodipteryx*.
8. Giant Swallows—*Nyctibius*.

Fourth Group—Cave Swallows—(*Steatornithes*)—One Family.

1. Guacharo—*Steatornis caripensis*.

Fifth Group—Owl Swallows—(*Podargi*)—Three Families.

1. Dwarf Owl Swallows—*Aegothehes*.
2. Owl Swallows—*Podargus*.
3. Frog-mouth Swallows—*Batrachostomus*.

SIXTH ORDER

SINGING BIRDS—(*Oscines*)—Thirty Groups.

First Group—Butcher-birds. Shrikes—(*Lanii*)—Three Families.

1. Preying Butcher-birds—*Lanius*.
2. Knockbirds—*Enneoctonus*.
3. Big-head Shrikes—*Falcunculus*.

Second Group—Bush Shrikes—(*Malaconoti*)—Three Families.

1. Flute Shrikes—*Laniarius*.
2. Cap Shrikes—*Telephonus*.
3. Helmet Shrikes—*Prionops*.

Third Group—Raven Shrikes—(*Thamnophili*)—Two Families.

1. Crow Shrikes—*Cracticus*.
2. Bataras—*Thamnophilus*.

Fourth Group—Snapping Shrikes—(*Edolii*)—Three Families.

1. Snapping Shrikes—*Dicrurus*.
2. Drongo's—*Chaptia*.
3. Flag Drongo's—*Edolius*.

Fifth Group—Swallow Shrikes—(*Artami*)—One Family.

1. Swallow Shrike—*Artamus sordidus*.

Sixth Group—King Shrikes—(*Tyranni*)—Six Families.

1. Tyrants—*Tyrannus*.
2. Crying Tyrants—*Saurophagus*.
3. Fork Tyrants—*Milvulus*.
4. Crown Tyrants—*Megalophus*.
5. Fly Stilts—*Gubernetes*.
6. Cockerels—*Alectrurus*.

Seventh Group — Caterpillar Eaters — (*Campephaga*) — One Family.

1. Vermilion Birds—*Pericrocotus*.

Eighth Group—Fly Snappers—*Myiagra*)—Two Families.

1. Paradise Snappers—*Terpsiphone*.
2. Fan Tails—*Rhipidura*.

Ninth Group—Fly Catchers—(*Muscicapa*)—Three Families.

1. Fly Catchers—*Butalis*.
2. Mourning Fly Catchers—*Muscicapa*.
3. Dwarf Fly Catchers—*Erythrosterna*.

Tenth Group — Waxwings, Cedar-birds — (*Bombycilla*) — One Family.

1. Waxwings—*Bombycilla*.

Eleventh Group—Ornament-birds—(*Pipra*)—Three Families.

1. Cliff-birds—*Rupicola*.
2. Ornament-birds—*Pipra*.
3. Panther-birds—*Pardalotus*.

Twelfth Group—Crop-birds—(*Gymnoderi*)—Three Families.

1. Capuchin-birds—*Gymnocephalus*.
2. Bell-birds—*Cephalopterus*.
3. Bell-birds—*Chasmarhynchus*.

Thirteenth Group—Ground Singers—(*Humicola*)—Five Families.

1. Nightingales—*Luscinia*.
2. Tree Nightingales—*Aëdon*.
3. Blue-throats—*Cyanecula*.
4. Rubi Nightingales—*Calliope*.
5. Redbreasts—*Rubecula*.

Fourteenth Group—Chats—(*Monticola*)—Six Families.

1. Red Stars—*Ruticilla*.
2. Meadow Chats—*Pratincola*.
3. Stone Chats—*Saxicola*.
4. Running Chats—*Dromolaea*.
5. Rock Chats—*Petrocincla*.
6. Bush Chats—*Thamnolaea*.

Fifteenth Group—Thrushes—(*Turdi*)—Two Families.

1. Wood Thrushes—*Turdus.*
2. Merles—*Merula.*

Sixteenth Group—Mocking Thrushes—(*Mimus*)—Three Families.

1. Mocking Thrushes—*Mimus.*
2. Red Mockers—*Taxostoma.*
3. Cry Mockers—*Galeoscoptes.*

Seventeenth Group—Bustle Thrushes—(*Timalus*)—Four Families.

1. Gray Thrushes—*Pycnonotus.*
2. Tattling Thrushes—*Timalus.*
3. Thrushlings—*Crateropus.*
4. Laughing Thrushes—*Garrulax.*

Eighteenth Group—Water Thrushes—(*Cincli*)—One Family.

1. Water Ouzels—*Cinclus.*

Nineteenth Group—Pomp Thrushes—(*Pitta*)—One Family.

1. Pittas—*Pitta.*

Twentieth Group—Ant Thrushes—(*Myiothera*)—Three Families.

1. Ant Birds—*Pyriglena.*
2. Ant Kings—*Grallaria.*
3. Rail Slippers—*Pteroptochus.*

Twenty-first Group—Lyre-tails—(*Menura*)—One Family.

1. Lyre-tails—*Menura.*

Twenty-second Group—Grassmonks, Hedge Sparrows—(*Sylvia*) —Two Families.

1. Grassmonks—*Curruca.*
2. Bush Singers—*Pyrophthalmus.*

Twenty-third Group—Leaf Singers—(*Phylloscopi*)—Three Families.

1. Leaf Singers—*Phyllopneuste.*
2. Leaf Kings—*Reguloides.*
3. Bastard Nightingales—*Hypolais.*

Twenty-fourth Group—Bullrush Singers—(*Calamodyta*)—Three Families.

1. Reed Singers—*Acrocephalus.*
2. Bullrush Singers—*Calamodus.*
3. Cricket Singers—*Locustella.*

Twenty-fifth Group—Bush Singers—(*Drymoica*)—Three Families.

1. Reed-grass Singers—*Cisticola.*
2. Taylor-birds—*Orthotomus.*
3. Emu Slippers—*Stipiturus.*

Twenty-sixth Group—Slippers, Wrens—(*Troglodita*)—Three Families.

1. Wren Slippers—*Troglodytes.*
2. Rush Slippers—*Thryothorus.*
3. Flageolet-birds—*Cyphorhinus.*

Twenty-seventh Group—Pipers—(*Anthi*)—Three Families.

1. Pipers—*Anthus.*
2. Field Pipers—*Agrodroma.*
3. Silk Pipers—*Corydalla.*

Twenty-eighth Group — Stilts, Wag-tails — (*Motacilla*) — Five Families.

1. Stilts—*Motacilla.*
2. Water Stilts—*Calobates.*
3. Sheep Stilts—*Budytes.*
4. Wood Stilts—*Nemoricola.*
5. Swallow Stilts—*Enicurus.*

Twenty-ninth Group—Flue Birds—(*Accentores*)—Two Families.

1. Wood Flue Birds—*Tharrhaleus.*
2. Flue Larks—*Accentor.*

Thirtieth Group—Titmice—(*Pari*)—Six Families.

1. Golden Crowns—*Regulus.*
2. Marsh Tits—*Aegithalos.*
3. Reed Tits—*Panurus.*
4. Tail Tits—*Orites.*
5. Tit Kings—*Lophophanes.*
6. Wood Tits—*Parus.*

THIRD CLASS.

SEARCHERS—(*Investigatores*)—THREE ORDERS.

SEVENTH ORDER.

CLIMBING BIRDS—(*Scansores*)—Seventeen Groups.

First Group—Flower Birds—(*Certhiola*)—Two Families.

1. Bluebirds—*Coereba.*
2. Pit-pits—*Certhiola.*

Second Group—Honey Suckers—(*Nectarinia*)—Three Families.

1. Honey Suckers—*Hedydipna.*
2. Fire Honey Suckers—*Aethopyga.*
3. Flower Gleaners—*Cyrtostomus.*

Third Group—Plantain Runners—(*Arachnothera*)—Two Families.

1. Half Bills—*Hemignathus.*
2. Hang Birds—*Arachnocentra.*

Fourth Group—Brush Tongues—(*Meliphaga*)—Four Families.

1. Honey Eaters—*Myzomela.*
2. Ear Tufts—*Ptilotis.*
3. Flower Tongues—*Melithara.*
4. Monk Birds—*Tropidorhynchus.*

Fifth Group—Hoopoes—(*Upupa*)—Two Families.

1. Hoopoes—*Upupa.*
2. Tree Hoopoes—*Irrisor.*

Sixth Group—Tree Mounters—(*Anabata*)—One Family.

1. Bunch Nestlers—*Phacellodomus.*

Seventh Group—(Potter Birds—(*Furnaria*)—Three Families.

1. Oven Birds—*Furnarius.*
2. Ground Mud-wall Makers—*Geosita.*
3. Mount Bills—*Neuopa.*

Eighth Group—Nuthatches—(*Sitta*)—Two Families.

1. Nuthatches—*Sitta.*
2. Tree Cleavers—*Sittella.*

Ninth Group—Wall Climbers—(*Tichodroma*)—One Family.

1. Alps Wall Climbers—*Tichodroma.*

Tenth Group—Tree Climbers—(*Scansores*)—Three Families.

1. Creepers—*Certhia*.
2. Tree Choppers—*Xiphorhynchus*.
3. Woodpecker Tree Choppers—*Dendroplex*.

Eleventh Group—Woodpeckers—(*Picidæ*)—Two Families.

1. Black Woodpeckers—*Dryocopus*.
2. Giant Woodpeckers—*Campephilus*.

Twelfth Group—Jay Woodpeckers—(*Melanerpes*)—One Family.

1. Jay Woodpeckers—*Melanerpes*.

Thirteenth Group—Spotted Woodpeckers—(*Pici*)—Four Families.

1. Spotted Woodpeckers—*Picus*.
2. Medium Woodpeckers—*P. medius*.
3. Little Woodpeckers—*P. minor*.
4. Three-toed Woodpeckers—*Apternus*.

Fourteenth Group — Green Woodpeckers — (*Gecinus*) — One Family.

1. Green Woodpeckers—*Gecinus viridis*.

Fifteenth Group—Cuckoo Woodpeckers—(*Colaptes*)—Two Families.

1. Gold Woodpeckers—*Colaptes*.
2. Field Woodpeckers—*Geocolaptes*.

Sixteenth Group — Dwarf Woodpeckers — (*Picumnus*) — One Family.

1. Dwarf Woodpeckers—*Picumnus*.

Seventeenth Group—Wry Necks—(*Jyngine*)—One Family.

1. Wry Necks—*Jynx*.

EIGHTH ORDER.

Humming Birds, or Columns of Short-wings—Eleven Groups.

First Group—Giant Gnomes—(*Eutoxeres*)—Two Families.

1. Giant Colibris—*Patagona*.
2. Sword Bills—*Docimastes*.

Second Group—Gnomes—(*Polytmi*)—Two Families.

1. Hawk Noses—*Grypus*.
2. Eagle Bills—*Eutoxeres*.

Third Group—Sun Birds—(*Phaethornis*)—One Family.

1. Solitaire—*Phaethornis*.

Fourth Group—Mountain Nymphs—(*Oreotrochili*)—Three Families.

1. Chimborazo Birds—*Oreotrochilus*.
2. Sword Wingers—*Campylopterus*.
3. Curve Wingers—*Platystylopterus*.

Fifth Group—Jewel Birds—(*Hypophania*)—Two Families.

1. Topaz—*Topaza*.
2. Cap Colibris—*Aithurus*.

Sixth Group—Wood Nymphs—(*Lampornithes*)—Two Families.

1. Mango's—*Lampornis*.
2. Wood Nymphs—*Chrysolampis*.

Seventh Group—Flower Nymphs—(*Floringi*)—Two Families.

1. Flower Kissers—*Heliothrix*.
2. Flower Suckers—*Florisuga*.

Eighth Group—Fairies—(*Trochili*)—Three Families.

1. Colibris—*Trochilus*.
2. Amethyst Birds—*Calliphlox*.
3. Point Tails—*Calothorax*, or *Lucifer*.

Ninth Group—Elfs—(*Lophornithes*)—Four Families.

1. Coe Elfs—*Cephalolepis*.
2. Pomp Elfs—*Lophornis*.
3. King Elfs—*Bellatrix*.
4. Tuil Elfs—*Heliactinus*.

Tenth Group—Sylphs—(*Lesbia*)—Two Families.

1. Banner Sylphs—*Steganurus*.
2. Train Sylphs—*Sparganura*.

Eleventh Group—Mask Colibris—(*Microramphi*)—Two Families.

1. Thorn Bills—*Ramphomicron*.
2. Helmet Colibris—*Oxypogon*.

NINTH ORDER.

Light Bills—(*Levirostres*)—Nineteen Groups.

First Group—Bee Eaters—(*Meropes*)—Six Families.

1. Bee Eaters—*Merops*.
2. Bee Wolves—*Melittotheres*.
3. Speckled Bee Eaters—*Coccolarynx*.
4. Forked Bee Eaters—*Melittophagus*.
5. Ornamented Bee Eaters—*Cosmaerops*.
6. Night Bee Eaters—*Nyctiornis*.

Second Group—Rakes—(*Coracii*)—Two Families.

1. Blue Rakes—*Coracias*.
2. Rollers—*Eurystomus*.

Third Group—Saw Rakes—(*Prionites*)—One Family.

1. Saw Rakes—*Prionites*.

Fourth Group—Throat Birds—(*Eurylaimi*)—Three Families.

1. Trowel Bills—*Corydon*.
2. Horn Throats—*Eurylaimus*.
3. Rayas—*Psarisomus*.

Fifth Group—Flat Bills—(*Todi*)—One Family.

1. Flat Bills—*Todus*.

Sixth Group—King Fishers—(*Alcedines*)—Three Families.

1. King Fishers—*Alcedo*.
2. Stump King Fishers—*Ceyx*.
3. Throat King Fishers—*Ceryle*.

Seventh Group—Halcions—(*Halcyones*)—Six Families.

1. Tree Halcions—*Halcyon*.
2. Wood Halcions—*Todirhamphus*.
3. Blue Halcions—*Cyanalcyon*.
4. Giant Halcions—*Dacelo*, or *Dacelo*.
5. Paradise Halcions—*Tanysiptera*.
6. Sawyer—Halcions—*Syma*.

Eighth Group—Lazy Birds—(*Galbulithes*)—One Family.

1. Jacamars—*Galbula*.

Ninth Group—Bush Cuckoos—(*Buccones*)—Three Families.

1. Sleep Birds—*Nystalus*.
2. Trappists—*Monasa*.
3. Drunkards—*Chelidoptera*.

Tenth Group—Gnaw Bills—(*Trogones*)—Five Families.

1. Fire Surukus—*Harpactes*.
2. Flower Surukus—*Hapaloderma*.
3. Surukus—*Trogon*.
4. Tocoloros—*Prionotelus*.
5. Pomp Surukus—*Calurus*.

Eleventh Group—Cuckoo Birds—(*Cuculidæ*)—One Family.

1. Honey Cuckoon—*Indicator*.

Twelfth Group—Cuckoos—(*Cuculi*)—Five Families.

1. Cuckoos proper—*Cuculus*.
2. Jay Cuckoos—*Coccystes*.
3. Cucklings—*Eudynamys*.
4. Gold Cuckoos—*Chrysococcyx*.
5. Silky Birds—*Scythrops*.

Thirteenth Group—Bush Cuckoos—(*Phœnicophæi*)—One Family.

1. Sickle Cuckoon—*Zanclostomus*.

Fourteenth Group—Heel Cuckoos—(*Coccyges*)—Three Families.

1. Rain Cuckoos—*Coccygus*.
2. Lizzard Cuckoos—*Saurothera*.
3. Slender Cuckoon—*Pyrrhococcyx*.

Fifteenth Group—Maggot Eaters—(*Crotophagi*)—One Family.

1. Maggot Eaters—*Crotophaga*.

Sixteenth Group—Spur Cuckoos—(*Centropodes*)—Three Families.

1. Spur Feet—*Centropus*.
2. Crow Pheasants—*Centrococcyx*.
3. Pheasant-spur Cuckoos—*Polophilus*.

Seventeenth Group—Barb Birds—(*Capitones*)—Three Families.

1. Pomp Barb Birds—*Trachyphonus*.
2. Gold Barb Birds—*Xantholæma*.
3. Toucan Barb Birds—*Tetragonops*.

Eighteenth Group—Toucans—(*Ramphasti*)—Two Families.

1. Aracaris—*Pteroglossus*.
2. Toucans—*Ramphastus*.

Nineteenth Group—Horn Birds—(*Bucerotes*)—Four Families.

1. Smooth-horn Birds—*Rhynchaceros*.
2. Double-horn Birds—*Dichoceros*.
3. Wrinkled-horn Birds—*Rhyticeros*.
4. Horn Ravens—*Buceros*.

FOURTH CLASS.

RUNNERS—(*Cursores*)—Four Orders.

TENTH ORDER.

Cooing Birds—(*Gyratores*)—Ten Groups.

First Group—Fruit Doves—(*Trerones*)—One Family.

1. Parrot Doves—*Phalacrotreron*.

Second Group—Doves—(*Columbæ*)—Two Families.

1. Ringed Doves—*Palumbus*.
2. Hole Doves—*Columba*.

Third Group—Cuckoo Doves—(*Macropygæ*)—One Family.

1. Wandering Pigeons—*Ectopistes*.

Fourth Group—Turtle Doves—(*Turtures*)—Three Families.

1. Turtle Doves proper—*Turtur*.
2. Laughing Doves—*Streptopelcia*.
3. Dwarf Doves—*Chalcopelcia*.

Fifth Group—Rail Doves—(*Zenaidæ*)—Four Families.

1. Trill Doves—*Melopelcia*.
2. Sparrow Doves—*Pyrgotænas*.
3. Sparrowhawk Doves—*Geopelcia*.
4. Wedge-tail Doves—*Stictopelcia*.

Sixth Group—Running Doves—(*Geotrygones*)—One Family.

1. Partridge Doves—*Sternænas*.

Seventh Group—Lustre Doves—(*Phapes*)—Four Families.

1. Tuft Doves—*Ocyphaps*.
2. Glimmer Doves—*Phaps*.
3. Quail Doves—*Geophaps*.
4. White-flesh Doves—*Leucosarcia*.

Eighth Group—Mane Doves—(*Callœnas*)—One Family.

1. Mane Doves—*Callœnas*.

Ninth Group—Crown Doves—(*Gouræ*)—One Family.

1. Crown and Fan Doves—*Goura*.

Tenth Group—Tooth Doves—(*Didunculi*)—One Family.

1. Tooth Doves—*Didunculus*.

ELEVENTH ORDER.

SCRATCHING BIRDS—(*Rasores*)—Seventeen Groups.

First Group—Flight Hens—(*Pteroclæ*)—Two Families.

1. Flight Hens—*Pterocles*.
2. Scepps or Heath Hens—*Syrrhaptes*.

Second Group—Rough-leg Hens—(*Tetraones*)—Five Families.

1. Ure Hens—*Tetrao*.
2. Playing Hens—*Lyrurus*.
3. Hazel Hens—*Bonasia*.
4. Prairie Hens—*Cupidonia*.
5. Snow Hens—*Lagopus*.

Third Group—Field Hens—(*Perdices*)—Five Families.

1. Rock Hens—*Tetraogallus*.
2. Red Hens—*Caccabis*.
3. Field Hens—*Perdix*.
4. Francolins—*Francolinus*.
5. Bare-neck Hens—*Pternistes*.

Fourth Group—Tree Hens—(*Odontophors*)—Three Families.

1. Tree Hens—*Odontophorus*.
2. Tree Quails—*Ortyx*.
3. Tuft Quails—*Lophortyx*.

Fifth Group—Quails—(*Coturnices*)—Two Families.

1. Quails—*Coturnix*.
2. Dwarf Quails—*Excalfactoria*.

Sixth Group—Running Hens—(*Turnaces*)—Two Families.

1. Running Hens—*Turnix*.
2. Bustard Quails—*Pedionomus*.

Seventh Group—Pheasant Birds—(*Phasianidæ*)—Two Families.

1. Pomp Hens—*Lophophorus*.
2. Horn Pheasants—*Ceriornis*.

Pl. XIII.

measured full nine feet, and were extremely thin. The crop or craw was of proportionate size, and the stomach large, resembling an oblong pouch. Both crop and stomach contained half-digested fish. The heart and lungs were large and strong. There was no muscular gizzard. The female bird is about two inches longer than the male. The upper portion of her head is less white than that of the male, and her breast is marked with brown streaks.

PLATE XVII.

The Cinereous Coot. (*Fulica Americana.*)

This species was formerly, by some ornithologists, classed among the Natatores, or swimming birds proper; but its form, the compressed body, and especially its mode of living, designate it clearly as a connecting link between the Gallinules and the swimming birds. It has a very strong resemblance, in the formation of its whole body, to the Gallinules, except that its feet are lobed.

The Cinereous Coot usually makes its appearance in the State of Ohio about the middle of April, stays the whole summer, and leaves for the South when the rushes are destroyed by severe frosts.

This bird is found almost everywhere in Europe, but is represented in the southern parts by a related kind. It has been found in middle Asia, and in its winter-quarters, in the interior of Africa. It is probable, however, that one or the other observer may have intermixed the different related kinds, not having taken the trouble of a close examination. In Great Britain it is said to be found at all seasons, and does not seem to migrate to other countries, but merely changes its station in autumn from the lesser pools or loughs, where their young are reared, to the larger lakes, where these birds assemble in winter in large flocks. They are also found in Germany. They avoid rivers and brooks as well as the sea, and prefer still waters, whose borders are overgrown with rushes and reeds.

They are consequently most numerous in the marshes of the larger lakes, and on the larger ponds. The time of their appearance in the spring depends chiefly, it seems, on the melting of the snow and ice. They remain in the same place during the whole summer, and in autumn begin to wander, assembling sometimes in immense flocks on the larger sheets of water, whence they migrate to the South, usually in the latter part of October and in November.

The Coot is oftener seen on the water than on land, but frequents the latter, especially during midday, to take a rest, and to clean and put its plumage in order. Though the feet of the Coot are rather awkwardly constructed for running, it runs tolerably well on the ground; but spends by far the greater part of its life in swimming. Its feet are excellent rudders, for whist their swimming lobes are lacking in breadth, is made up by the length of the toes. The Coot is also an expert diver, and contents the palm. In this respect, with many real swimming birds. It dives to considerable depths, and swims, with the help of its wings, great distances under water. To escape danger, it always sinks itself in deep water. Before it rises for a flight, it flutters for a great distance over the surface, making the water so violently with its feet that the noise of the splashing can be heard at a great distance.

The Coot is very loquacious, chattering to its companions almost incessantly. Its voice is a shrill "Kara," and the shrillness, in time so anger, is doubled or even trebled. It also utters a short, hard "Pētz," and at times a hollow guttural sound. It is a very sociable bird among its own kind, except in the breeding season, when each pair always strive to keep a certain district for themselves, into which they never suffer any other birds to come. Even in their winter-quarters, Coots do not like to see other swimming birds, and make it a special point to drive away Ducks.

Aquatic insects and their larvæ, worms and small snails, and several kinds of vegetable matter, which they find in the water, form the principal food of Cinereous Coots. They pick up their food in swimming and diving, either from the surface, or by diving after it to the bottom. Some Coots, kept in captivity, lived for a whole winter exclusively on grain, and although they were occasionally fed with small minnows, which they readily ate, they seemed to prefer the grain. Whenever the Coot has settled on the smaller ponds or swamps it begins to build its nest, which is formed in the rushes near the water's edge. It is built on the trampled down stocks of weeds and rushes, and is composed of the dry stocks of the same. The upper layers and the interior consist of a little finer material, such as the finer weeds, dry grass, and fibers. The female lays, in the latter part of May, from seven to twelve eggs, rather large in proportion to the size of the bird, having a fine but hard shell, of a yellowish brown color, sprinkled over with dark ash colored and blackish brown dots, chiefly on the large end. The eggs are hatched in about twenty or twenty-one days. As soon as the young quit the shell and are dry, they plunge into the water, and dive and swim with the greatest ease, but always cluster again about the mother, taking shelter under her wings, while the male warns and protects them from danger. For a considerable time they return nightly to their nest; but gradually they separate more and more from the parents. Long before they are fully fledged, they become independent of parental care.

The female Coot frequently breeds twice in a season, but may be called lucky if she raises one-half of the young she hatches. Great havoc is made among them, before they have learned by experience to defend themselves, by the Marsh Hawk and other kinds of the Hawk tribe, as well as by turtles.

A Coot is found in Europe, the *Fulica âtra*, resembling the American, though differing from it in having for bill and frontal plate perfectly white, while on the American Coot the frontal plate is always of a bright chestnut color. The Coot's gizzard is strong and muscular, like that of a common hen. The male and female are colored alike, except that the black on the head and neck of the female is less brilliant. The flesh of the Coot, even that of the young, makes an unsavory dish for the table.

PLATE XVIII.

The Pileated Woodpecker. (*Dryocopus—Picus Pileatus.*)

Fig. 2.

This Woodpecker, second only in size to any other, is a true American bird, and may be called the chief of all northern Woodpeckers. His range extends from Upper Canada, all over the United States, to the Gulf of Mexico. He abounds most in the North, in forests of tall trees, particularly in the neighborhood of large rivers, where he is noted for his loud cries, especially before wet weather. At such times he flies, restless and uneasy, from tree to tree, making the forest echo with his outcries. In the State of Ohio, and generally in all the Northern States, he is called the **Black** Woodcock; in the Southern States they call him the Logcock. Every old trunk in the forest where he resides, bears more or less the marks of his chisel-like bill. Whenever he finds a tree beginning to decay, he subjects it to a close examination, in order to find out the cause, going round and round it, and pulling the bark off in strips often several feet long, laboring with astonishing skill and activity. He has frequently been seen to strip the bark from a dead pine tree, eight or ten feet down, in less than fifteen minutes. Whatever he is doing, whether climbing, stripping off bark, or digging, he seems to be always in great haste. He is extremely watchful and shy, and is consequently difficult to kill. He clings closely to the tree after having received his mortal wound, and does not even quit his hold with his last breath. If shot at on the wing, and only one wing is broken, as soon as he drops to the ground he makes for the nearest tree and climbs on it high enough to be out of reach. When wounded, and lying on the ground, he strikes with great fierceness at the hand stretched out to seize him. He is one of the few birds that are never content when caged or confined.

This bird is now in one part of his district and then directly in another part, roaming through the whole of it in an incredibly short time. In the course of a few minutes, his cries are heard in different places, remote from each other. He utters three principal cries—two in flying, and the other when sitting or climbing; the former sounding like "Kerr, Kerr," and "Kleeck, Kleeck;" the latter like "Kleh," lengthened out and penetrating, or like "Kleha, Kleha." Besides these cries he has several others, which he utters for the most part when near his nest. His flight is different from that of other Woodpeckers. He does not, like them, fly by starts, or in alternately ascending and descending lines, but wavelike forward in a straight direction, spreading his wings far apart and striking the air hard, so that the points of the larger quills appear to be bending upward, causing his flight to resemble that of the Jay. It is, however, more gentle than that of the other Woodpeckers, and seems to require less exertion. The distinct whirr which we hear in their flight, we do not hear in his. Although he seems averse to long flights, he has been observed flying directly forward, without stopping, for the distance of about half a mile. He hops rather awkwardly on the ground, where he is frequently seen examining the ant-hills in quest of the larvæ or eggs, of which he seems to be extremely fond. In climbing, and boring with his chisel-like bill, he is very expert. When he climbs, he puts both feet forward like all other Woodpeckers. He may therefore be said to hop up the trees, and this he does with great force, so that one can distinctly hear his claws striking into the bark. While climbing he keeps his breast away from the trunk, bending his neck backward.

His food consists of ants and their larvæ, which he picks up with his sticky tongue. He also devours the larvæ of beetles found in pine forests, and to get at them he chisels large holes in the trees. The mating season of these birds is in April, early or late, according to the season. The male at that period flies after the female, crying aloud, and coming up to her, or becoming tired of flying after her, he alights on the withered top of a tree and begins to drum. He chooses on the tree a place where the beating of his bill will resound the loudest. Pressing his tail hard against a dead limb, he raps so quickly and forcibly upon it with his bill, that the noise made sounds like a continued "Er-r-r-r-r-r." The rapid motion of the red top on his head appears like a glowing spark on the end of a burning stick, moved quickly to and fro. The female makes her appearance after the drumming, or sometimes answers by quickly repeated "Kluck, Kluck, Kluck." The male also keeps up his drumming while the female is sitting on the eggs.

For their nests these birds seek a decayed or hollow tree, choosing a knot-hole for the entrance. This hole is widened by the female, so as to make it sufficiently large for going in and out with ease. The inner part of the tree is then hollowed out with peculiar dexterity. This process seems to be very difficult for the female, as there is not room enough for working with her bill. The sounds made are very dull, the chips small, and the work progresses slowly; but as soon as she has gained more room, she is enabled to dig out larger chips, and the work goes on more rapidly. Chips have been found under a tree where she was at work, from four to five inches long and half an inch in breadth and thickness. The female only works in the forenoon, going out in the afternoon after food. After laboring hard from ten to sixteen days, she has the nesting-place prepared. It is from fifteen to twenty inches deep and from eight to ten inches in diameter, the sides being very smooth, and the bottom ball-shaped and covered with fine chips. On these chips the female lays three, four, and sometimes five eggs, which are rather small and of a brilliant white color, looking like

enamel. The nest is usually built high up on a tree, generally on a pine tree.

The same nest is used for several years, but is usually cleaned out and enlarged. The male assists the female in hatching, the female sitting on the eggs during the night and the early morning hours. The newly hatched young are ill-shaped, being sparingly covered on the upper part of the body with a grayish black down, and the head being very large and the bill thick and clumsy. The parents seem to be very fond of their young, and utter mournful sounds when any one approaches the nest, and risk even their own lives in defense of their brood. The young are fed from the crops of the old birds, and their food consists chiefly of the so-called eggs of the black ant. If not disturbed, they remain in the nest until perfectly fledged; but before that time they often climb up to the entrance and take a look at the outside world.

The Hairy Woodpecker. (Picus Villosus.)

Fig. 2

This species may be regarded as a true type of the Woodpeckers (Pici). They are found almost exclusively on the trunks of trees, and are seldom seen on the ground. They are resident birds, and rarely missed in the orchards, where they are always busily engaged in boring apple trees, eagerly hunting for insects, their eggs or larvæ, in old withered stumps, rotten branches, and crevices of the bark. They inhabit North America from Hudson's Bay to the Carolinas and Georgia. In May, this Woodpecker retires to the groves and deeper forests with his mate to breed, though they frequently choose the orchard for that purpose, and select a suitable apple or pear tree. They seek a branch already hollow, or dig out an opening for their nest. The nest has been found more than four feet from the mouth of the hole. They dig first horizontally, if in the trunk of a tree, for six or seven inches, and then obliquely down for twelve or fifteen inches, carrying the chips out with their bills or scraping them out with their feet. A nest is now made with fine chips at the bottom of the hole. The female lays from four to five bluish white eggs and hatches them out in June. Their residence in summer is limited to a comparatively small extent of country; but in the fall and winter they roam about in a larger district, and usually in company with Nuthatches, Creepers, Titmice, and Golden-crested Wrens. In summer they never suffer another bird of their kind to come within their district. They make their appearance in a moment, as soon as they hear a knocking resembling that of another Woodpecker. In their roamings they fly chiefly from tree to tree, avoiding large open spaces.

These Woodpeckers are lively, active, and daring. Their contrasted colors make them look beautiful, even when seen from a distance, and especially when they are flying. It is a fine sight when on a clear, sunny day they chase each other from tree to tree, or climb swiftly up in the sunshine on the branches or the trunk of a tree, or when they bask in the sunlight on the tops of high trees, or on a withered limb execute their playful drumming. They are almost constantly in motion, and enliven the forests, especially the dark pine woods, in a most agreeable manner. Their flight is swift and produces a humming; but it is usually not far extended. They rarely come down to the ground, but when on it, hop about with considerable skill. They prefer to sit on the tops of the trees, repeating their "pick, pick, pick," or "kick, kick, kick." Their sleeping-places, like those of all Woodpeckers, are hollow trees, and to these they retire when wounded. Such is their conduct toward their own kind and toward other birds that they can not be called sociable. They can be easily deceived by imitating their drumming, especially in the spring-time, as at that time, besides their desire for food, jealousy is brought into play. In summer, when thus deceived, they appear close before you, climbing about on all the branches to get a sight of the supposed rivals or intruders; on such occasions both the male and female make their appearance.

Their food consists of different kinds of insects, their eggs and larvæ, and also of nuts and berries. It is principally gathered from trees. For their young, they chiefly pick up small caterpillars. They are very useful in forests and orchards, as they destroy the insects that infest the trees. Frequently, after a few hard raps with their bills on a small limb, they run round to the opposite side to pick up the insects that the jarring has started out. The male and female alternately sit on the eggs, and the young break out of the shell in fourteen or sixteen days. They are at first helpless and deformed, but are most tenderly taken care of by their parents, who, when there is any seeming danger, wail piteously and never leave the nest. For a long time after the young are fully fledged, they are guarded and fed by the parents until perfectly able to find their own food and take care of themselves. The male and female birds are alike in color, except that the female lacks the red on the hind head, and the white below is tinged with brown. The name of Hairy Woodpecker is doubtless bestowed upon this bird on account of the white lateral spot on the back, composed of loose feathers resembling hair. This bird usually utters a loud tremulous cry in starting off, and when alighting. When mortally wounded it will hang by the claws, even of a single foot, while a spark of life remains.

PLATE XIX.

The Clapper Rail. (*Rallus Crepitans.*)

Fig. 1

The Clapper Rail, designated by different names, such as the Mud Hen, Meadow Clapper, Big Rail, and several others, is a well-known and very numerous species, inhabiting the whole Atlantic coast from Florida to New England, and probably still more northward. Although they chiefly inhabit the salt-marshes, these birds are occasionally found on the swampy shores and tide waters of our large rivers, as well as on the lakes. They, as well as other rails, are birds of passage, arriving on the coasts the latter part of April, and leaving late in September. They have been observed in great numbers at the mouth of the Savannah river, in the months of January and February, and it is therefore very probable that some of them winter in the marshes of Georgia and Florida. They are often heard to cry while on their spring migrations, pretty high up in the air, generally a little before day-break. The shores, within the beach, consisting of large extents of flat marsh overgrown with rank and reedy grass or rushes, occasionally overflowed by the sea, by which they are cut into numberless small islands with narrow inlets, are the favorite breeding-places of the Clapper Rails, which are found there in double the number of all other marsh-birds.

The arrival of the Clapper Rail is announced by his loud, harsh, and incessant crackling, which bears a strong resemblance to that of the Guinea-fowl. It is generally heard during the night, and is greatest before a storm. Toward the middle of May the Clapper Rails begin to construct their nests and lay their eggs. They drop their first egg in a cavity lined with only a little dry grass, to which is gradually added, as the number of eggs increases, more and more grass, so that by the time the number of eggs reaches the full complement, usually nine or ten, the nest has attained a height of ten or fourteen inches. The reason for building the nest so high is doubtless to secure them from the rising of the tides. The large rank marsh-grass is skilfully arched over the nest, and knit at the top, in order to conceal the nest from view, and afford shelter against heavy rains; but instead of concealing the nest, it enables the experienced egg-hunter to find it more easily, for he can distinguish the spot when it is at a distance of from thirty to forty yards, although an unpractised eye would not be able to discern it at all. The eggs

are of a pale clay color, sprinkled over with numerous small spots or dots of a dark red. They measure fully an inch and a half in length by one inch in breadth, and are obtuse at the small end. They are considered exquisite food, far surpassing the eggs of the domestic hen. The proper time for collecting these eggs is about the beginning of June. The nests are so abundant, and some persons are so skilled in finding them, that sometimes from forty to fifty dozen are collected in one day by a single individual.

The Crows, Minks, and other animals hunt their eggs and destroy, not only a great number of them, but many of the birds also. Heaps of bones, feathers, wings, and eggs of the Clapper Rail are often found near the holes of Minks, by which these animals themselves are in turn detected, driven out, and killed.

The poor Clapper Rails are subjected to another calamity of a more serious and disastrous nature. It happens sometimes, after the greater part of the eggs are laid, that a violent northeast storm arises, and drives the sea into the bay, overflowing the marshes, and destroying all the nests and eggs. Besides, vast numbers of the birds perish, as the water rushes in suddenly, and the birds being entangled are unable to extricate themselves in time to escape drowning. Hundreds of these birds may be seen at such times floating over the marshes in great distress, a few only escaping to the mainland. On such occasions great numbers may sometimes be seen in a single meadow, bewildered and not trying to conceal themselves; while the bodies of female birds that perished in their nests are washed to the shore, with scarcely a male among the dead bodies. After such an occurrence the birds go to work again as soon as the water subsides, and in about a fortnight the nests and eggs are about as numerous as they were before the calamity. Instances have occurred when such a disaster happened twice in a breeding-season, and yet the Clapper Rails were not discouraged, but commenced building nests and laying eggs for the third time.

The young of the Clapper Rails bear a strong resemblance to the young of the Virginia Rails, although they are somewhat larger. They are covered, as well as the young Virginia Rails, with a soft black down, but differ from the latter in having a whitish spot on the auriculars, and a whitish streak along each side of the breast, belly, and fore part of the thigh. The legs are of a blackish slate color. These birds have a little white protuberance near the tip of the bill, and they are also whitish around the nostrils. They run with the greatest facility among the long grass and reeds, and can only be caught with great difficulty. Several young Clapper Rails caught in the marshes in New Jersey, about the middle of July, corresponded with the above description, the males and females being marked alike. The extreme nervous vigor of its limbs, and its compressed body, which enables it to run among the grass, reeds, and rushes with the greatest rapidity, seemed to be the only means of defense of this bird. Almost everywhere among the salt-marshes are covered passages under the flat and matted grass, through which the Rail makes its way like a rat, without being noticed. From nearly every nest runs one or more of these covered roads to the water's edge, by which the birds can escape unseen. If closely pursued, the Rail will dive and swim to the other side of the pond or inlet, rising and disappearing with celerity and in silence. In smooth water the Rail swims tolerably well, but not fast; he sits rather high in the water with the neck erect, striking out with his legs with great rapidity. On shore, he runs with the neck extended, frequently flirting up his erect tail, and running on smooth ground nearly as fast as a man.

These birds are always very difficult to catch on land even when their wings are broken. They can remain under water four or five minutes, clinging closely to the roots of rushes with the head bent downward. Their flight resembles that of a Duck. They generally fly low above the ground, with the neck extended, and with great velocity; but like all the Rail tribe they have a dislike to take wing, and whenever you traverse the marshes and accidentally start one Clapper Rail, you may be sure that there are hundreds of these birds, which, if hunted by a dog, will lead him through numberless labyrinths, and only flush when he is just at the point of seizing them.

The male and female Clapper Rails are colored nearly alike; but the young birds of the first year differ somewhat from them in color. The upper parts of these young birds are of a brownish olive streaked with a pale slate color: the wings are of a pale brown olive; the chin and throat white; the breast pale ash colored, and tinged with yellowish brown: the legs and feet are of a light horn color. These birds are never found at a great distance from the lakes; in large flocks in the interior part of the country, on the lakes they are frequently found, but never in great numbers. The Clapper Rail feeds chiefly on small shelled fish, especially on those of that town of small found so abundant in the marshes; but he also eats worms, which he digs out of the mud, and for which work his bill is admirably adapted. He also feeds on small crabs.

In the month of October, Clapper Rails migrate to the South, never in flocks, but singly or in pairs, flying high up in the air. None of them remain North during the winter, though one of them was killed in the Reservoir, about thirty-three miles north-east from Columbus, Ohio, in the latter part of November: but on a close inspection, it was found that the bird had been crippled.

The Belted Kingfisher. (*Alcedo Alcyon.*)

Fig. 4.

The Belted Kingfisher is an inhabitant of the shores and banks of all our fresh-water rivers from Hudson's Bay to Mexico. He seems to love running streams and falling waters, like the whole of his tribe. At such places, resting on an overhanging bough above a cascade, he will remain for hours, glancing around his piercing eyes in all directions, seeking to discern in the water below small minnows, which, as soon as seen, with a sudden circular plunge, executed with the velocity of an arrow shot from the bow, he sweeps from their element and swallows in an instant. The voice of the Belted Kingfisher resembles the sound of a child's rattle; it is sudden, harsh, and very loud, but in a certain degree softened by the murmuring of the brooks, or the sound of the cascades or brawling streams, among which he generally resides. He courses up and down the stream, along its different windings, at no great height above the water, sometimes poising himself by the rapid action of his wings, in the instant of some of the Hawk tribe, in order to pounce down into the water on some small fish, which he frequently misses. After such a miss he quietly settles with a dissatisfied look, on an old dead overhanging limb of a tree to shake off the water from his plumage and to reconnoitre again. Mill-dams are frequented by him, as the neighborhood usually abounds with small fish. Rapid flowing streams, with steep high banks of a clay or of a gravelly nature, are also his favorite places of resort, as in such steep and dry banks he usually digs a hole for his nest. This hole he digs with his bill and claws, extending it horizontally, sometimes to four or even six feet, and about half a yard below the surface, with a small cavity at the bottom for its nest. This is composed of a few fibres, a few dried fish-bones, and a little dry grass. The female lays five pure white eggs, comparatively of rather a large size. The young are hatched about the beginning of June; but this time differs according to the climate of the country where the breeding takes place. In the southern parts of the United States, the female Kingfisher has been found sitting on her eggs as early as the beginning of April, while in Ohio the Kingfisher nests, with the birds sitting on the eggs, are not usually found till toward the end of May. They occupy the same hole for several years as a breeding-place, and will not readily forsake it, even though it should be entered. There are accounts of people taking away the eggs of a Kingfisher, leaving one in the nest, and repeating this on their food collected twelve, or even eighteen eggs, the female always laying regularly one egg every

day. Such accounts being doubted, an experiment was made by taking from a nest-hole in the steep bank of the Connecticut river, a little below Middletown, Connecticut, the second egg laid; but instead of laying another egg, the birds abandoned the nest altogether. A similar experiment was tried in Ohio, with a like result.

In the Eastern and Western States, the Kingfisher generally remains until the commencement of the cold season, when he leaves for warmer regions, though he is occasionally seen in the Northern States in the middle of winter. He is found in the Southern States during nearly the whole winter. The Belted Kingfisher is like all the rest of the Kingfisher tribe, not much inclined to society, but is generally seen singly or in pairs, or in small groups of three or four. When coursing from one brook to river to another, or from one lake to another, which the Kingfisher frequently does, he passes over cities or forests in a bee-line, not unfrequently for a distance of ten or twenty miles. At such times his motions consist of five or six flaps, followed by a glide without making any undulations like the Woodpecker. In May, 1830, on a little creek in Connecticut, called the Hockanum, a Belted Kingfisher was observed on the ground, flapping his wings and seemingly in great distress. On coming up to him the observer found that his bill was stuck fast in a large clam. He had probably seen the clam on the muddy bank of the creek, with the shell partly open, and, in the attempt to pull the clam out, the shell had closed upon his bill. The possessor of course liberated the poor bird, which kind act he acknowledged by biting his benefactor on the thumb, and by shaking his rattle at him most indignantly as he flew away.

PLATE XX.

The Ash-colored or Black-cap Hawk. (*Falco atricapillus.*)

This beautiful Hawk has been confounded by many Ornithologists with the Goose Hawk of Europe; but there is such a difference between them that it is really wonderful how the two birds could be supposed to be identical. The greatest difference between these birds is in the markings of their breast and under parts, and this difference is so distinct as at once to strike the beholder. On our Hawk the under parts are of a uniform pale grayish white, each feather having in the center a black streak; this extends to the feathers to the center of the belly, after which the streak is hardly any more visible: besides this, every feather is marked transversely with fine, irregular zigzag bars of dark gray. In the European bird, each feather on the breast and lower parts is marked with a dark shaft, not exceeding its own breadth, and has besides two decided transverse bars, giving the bird, at a first glance, a very different aspect from the American Hawk. The upper parts of the latter are of a blue shade, and the markings of the head are darker and more decided. Some Ornithologists have classed this Hawk with the genus Astur, while others make it a sub-genus of Accipiter, in which the Sparrow Hawk and lesser species have been placed. Although there is some difference in the formation of the tarsi, the habits and instances in general are quite similar. The Broad-winged Hawk (Astur Pennsylvanicus) is an example of the one, and our Hawk that of the other.

The Black-capped Hawk is very spirited, and his general form and aspect denote great strength; his legs are very strong, and his claws rather large in proportion. The claws of the inner toes being as large as those of the great toes; his wings are short and too short, showing, when expanded, a slender bit more considerable to a smooth cutting flight, which is greatly aided by the lengthened tail. His reverse stories are forests or wooded country, where he can be seen hunting his prey about the skirts of the woods. In such places he builds his nest, usually on a high

Pl. 1.

Eighth Group—Comb Hens—(*Galli*)—One Family.
1. Wild Hens—*Gallus*.

Ninth Group—Pheasants—(*Phasiani*)—Seven Families.
1. Pheasant Hens—*Euplocamus*.
2. Silver Pheasants—*Nycthemerus*.
3. Noble Pheasants—*Phasianus*.
4. Collar Pheasants—*Thaumalea*.
5. Ear Pheasants—*Crossoptilon*.
6. Argus Pheasants—*Argus*.
7. Glimmer Pheasants—*Polyplectron*.

Tenth Group—Peacocks—(*Pavones*)—One Family.
1. Peacocks—*Pavo*.

Eleventh Group—Pearl Hens—(*Numida*)—Three Families.
1. King Pearl Hens—*Acryllium*.
2. Tuft Pearl Hens—*Guttera*.
3. Pearl Hens—*Numida*.

Twelfth Group—Turkeys—(*Meleagrides*)—One Family.
1. Turkeys—*Meleagris*.

Thirteenth Group—Brush Turkeys—(*Tallegalli*)—Three Families.
1. Thick-bill Hens—*Catheturus*.
2. Maleos—*Megacephalon*.
3. Dove Brush Turkeys—*Leipoa*.

Fourteenth Group—Great-foot Hens—(*Megapodii*)—One Family.
1. Great-foots—*Megapodius*.

Fifteenth Group—Hokko Birds—(*Cracida*)—Two Families.
1. Hokkos—*Crax*.
2. Mountain Hokkos—*Oreophasis*.

Sixteenth Group—Shaku Hens—(*Penelopa*)—Two Families.
1. Guan Hens—*Penelope*.
2. Tuft Guan Hens—*Opisthocomus*.

Seventeenth Group—Buttocks Hens—(*Crypturida*)—Four Families.
1. Injambus—*Crypturus*.
2. Great Buttocks Hens—*Rhynchotus*.
3. Dwarf Buttocks Hens—*Nothura*.
4. Macucas—*Trachypelmus*.

TWELFTH ORDER.

SHORT-WINGLERS—(*Brevipennes*)—Three Groups.

First Group—Ostriches—(*Struthiones*)—Three Families.
1. Ostriches—*Struthio*.
2. Nandu's—*Rhea*.
3. Emu's—*Dromaeus*.

Second Group—Cassowaries—(*Cassarii*)—One Family.
1. Cassowaries—*Casuarius*.

Third Group—Snipe Ostriches—(*Apteriges*)—One Family.
1. Snipe Ostriches—*Apteryx*.

THIRTEENTH ORDER.

STILT BIRDS—(*Grallatores*)—Twenty-nine Groups.

First Group—Bustards—(*Otides*)—Four Families.
1. Bustards—*Otis*.
2. Dwarf Bustards—*Tetrax*.
3. Hubaras—*Hubara*.
4. Ornament Bustards—*Sypheotides*.

Second Group—Run Birds—(*Tachydromi*)—Two Families.
1. Desert Runners—*Cursorius*.
2. Crocodile Watchers—*Hyas*.

Third Group—Fallow Swallows—(*Trochelia*)—One Family.
1. Fallow Swallows—*Glareola*.

Fourth Group—Thick-feet—(*Oedicnemi*)—One Family.
1. Thick-feet—*Oedicnemus*.

Fifth Group—Rain Pipers, or Plovers—(*Charadrii*)—Three Families.
1. Golden Plovers—*Charadrius*.
2. Alps Plovers—*Eudromias*.
3. Shore Plovers—*Ægialites*.

Sixth Group—Lapwings—(*Vanelli*)—Three Families.
1. Lapwings—*Vanellus*.
2. Spur Lapwings—*Hoplopterus*.
3. Lobe Lapwings—*Sarciophorus*.

Seventh Group—Turnstones—(*Strepsilati*)—One Family.
1. Turnstones—*Strepsilas*.

Eighth Group—Oyster-catchers—(*Hæmatopi*)—One Family.
1. Oyster-catchers—*Hæmatopus*.

Ninth Group—Snipe Birds—(*Limicola*)—Three Families.
1. Snipes—*Scolopax*.
2. Swamp Snipes—*Gallinago*.
3. Moor Snipes—*Philellanas*.

Tenth Group—Strand Runners, or Sand Pipers—(*Tringa*)—Five Families.
1. Swamp Runners—*Limicola*.
2. Sanderlings—*Calidris*.
3. Mud Runners—*Pelidna*.
4. Dwarf Strand Runners—*Actodroma*.
5. Wrestling Runners—*Philomachus*.

Eleventh Group—Water Steppers—(*Phalaropi*)—Two Families.
1. Odin Hens—*Lobipes*.
2. Water Steppers—*Phalaropus*.

Twelfth Group—Water Runners—(*Totani*)—Four Families.
1. Strand Pipers—*Actitis*.
2. Rain Snipes—*Glottis*.
3. Shore Snipes—*Limosa*.
4. Stilt Runners—*Hypsibates*.

Thirteenth Group—Sword-bills, or Avosets—(*Recurvirostra*)—One Family.
1. Avosets, or Sword-bills—*Recurvirostra*.

Fourteenth Group—Curlews, or Fallow Birds—(*Numenii*)—One Family.
1. Curlews—*Numenius*.

Fifteenth Group—Ibis's—(*Ibides*)—Three Families.
1. Sicklers—*Falcinellus*.
2. Scarlet Ibis's—*Ibis*.
3. Ibis—*Threskiornis*.

Sixteenth Group—Spoon-bill Herons—(*Plataea*)—One Family.
1. Spoon-bills—*Platalea*.

Seventeenth Group—Boat-bills—(*Cancromata*)—Two Families.
1. Shoe-bills—*Balaeniceps*.
2. Savakus—*Cancroma*.

Eighteenth Group—Shadow Birds—(Scopi)—One Family.

1. Shadow Birds—*Scopus.*

Nineteenth Group—Storks—(Ciconiæ)—Six Families.

1. Nevermiddatus—*Tantalus.*
2. Storks—*Ciconia.*
3. Simbils—*Sphenorhynchus.*
4. Giant Storks—*Mycteria.*
5. Coop Storks—*Leptoptilus.*
6. Gaphstus—*Anastomus.*

Twentieth Group—Herons—(Ardea)—Six Families.

1. Fishing Herons—*Ardea.*
2. Ornament Herons—*Herodias.*
3. Cow Herons—*Bubulcus.*
4. Night Herons—*Nycticorax.*
5. Dwarf Herons—*Ardetta.*
6. Bitterns—*Botaurus.*

Twenty-first Group—Sun Herons—(Eurypyga)—One Family.

1. Sun Herons—*Eurypyga.*

Twenty-second Group—Cranes—(Grues)—Two Families.

1. Cranes—*Grus.*
2. Virgin Cranes, or Demoiselles—*Anthropoides.*

Twenty-third Group—Crowned Cranes—(Balearica)—One Family.

1. Peacock Cranes—*Balearica.*

Twenty-fourth Group—Field Storks—(Arvicola)—Two Families.

1. Snake Storks—*Dicholophus.*
2. Trumpeter Birds—*Psophia.*

Twenty-fifth Group—Weapon Birds—(Palamedea)—Two Families.

1. Weapon Birds—*Palamedea.*
2. Tshajas—*Chauna.*

Twenty-sixth Group—Rails—(Ralli)—Four Families.

1. Snipe Rails—*Rhynchæa.*
2. Water Rails—*Rallus.*
3. Hen Rails—*Aramides.*
4. Corncrakes, or Meadow Rails—*Crex.*

Twenty-seventh Group—Leaf Pullets—(Parra)—Two Families.

1. Spurwings—*Parra.*
2. Water Pheasants—*Hydrophasianus.*

Twenty-eighth Group—Water Hens—(Gallinula)—Three Families.

1. Sultans Hens—*Porphyrio.*
2. Moor Hens—*Stagnicola.*
3. Coots—*Fulica.*

Twenty-ninth Group—Hemfeet—(Podoa)—One Family.

1. Diving Pullets—*Heliornis.*

FIFTH CLASS

SWIMMERS—(*Natatores*)—FOUR ORDERS.

FOURTEENTH ORDER

TOOTH-BILLS—(*Lamellirostres*)—Six Groups.

First Group—Stilt Swans—(Phœnicopteri)—One Family.

1. Flamingoes—*Phœnicopterus.*

Second Group—Swans—(Cygni)—Five Families.

1. Knob-bill Swan, Common Swan—*Cygnus olor.*
2. Song Swan—*C. musicus.*
3. Dwarf Swan—*C. Bewickii.*
4. Black-necked Swan—*C. nigricollis.*
5. Black Swan—*C. chenopsis atratus.*

Third Group—Geese—(Anseres)—Eight Families.

1. Spur Geese—*Plectropterus.*
2. Swan Geese—*Cygnopsis.*
3. Wild Geese—*Anser.*
4. Snow Geese—*Anser-Chen.*
5. Sea Geese—*Bernicla.*
6. Fox Geese—*Chenalopex.*
7. Dwarf Geese—*Nettapus.*
8. Hen Geese—*Cereopsis.*

Fourth Group—Swimming Ducks—(Anseres)—Seven Families.

1. Fox Ducks—*Casarca.*
2. Cave Ducks—*Vulpanser.*
3. Tree Ducks—*Dendrocygna.*
4. Speculum Ducks—*Anas.*
5. Ornament Ducks—*Aix.*
6. Spoon-bill Ducks—*Spatula.*
7. Musk Ducks—*Cairina.*

Fifth Group—Diving Ducks—(Fuligula)—Four Families.

1. Eider Ducks—*Somateria.*
2. Mourning Ducks—*Oidemia.*
3. Table Ducks—*Aythya.*
4. Rudder Ducks—*Erismatura.*

Sixth Group—Sawyers—(Mergi)—Two Families.

1. Dwarf Sawyers—*Mergellus.*
2. Tooth Sawyers—*Mergus.*

FIFTEENTH ORDER

SEA FLYERS—(*Longipennes*)—Eight Groups.

First Group—Sea Swallows, or Terns—(Sternæ)—Six Families.

1. Preying Sea Swallows—*Sylochelidon.*
2. Stream Swallows—*Sterna.*
3. Dwarf Sea Swallows—*Sternula.*
4. Water Swallows—*Hydrochelidon.*
5. Fairy Swallows—*Gygis.*
6. Noddy Tern—*Anous stolidus.*

Second Group—Scissor-bills—(Rhynchops)—One Family.

1. Scissor-bills—*Rhynchops.*

Third Group—Gulls—(Lari)—Five Families.

1. Fishing Gulls—*Larus.*
2. Ice-field Gulls—*Pagophila.*
3. Stump Gulls—*Rissa.*
4. Cap Gulls—*Chroicocephalus.*
5. Rose Gulls—*Rhodostethia.*

Fourth Group—Preying Gulls—(Lestris)—Two Families.

1. Preying Gulls—*Lestris.*
2. Parasite Gulls—*L. Stercorarius.*

Fifth Group—Albatrosses—(Diomedea)—One Family.

1. Albatross—*Diomedea.*

Sixth Group—Petrels, or Storm Birds—(*Procellariæ*)—Three Families.

1. Giant Petrels—*Procellaria Ossifraga*.
2. Ice Petrels—*Priocellaria*.
3. Duck Petrels—*Prion*.

Seventh Group—Storm Swallows—(*Oceanides*)—Two Families.

1. Storm Swallows—*Thalassidroma*.
2. Storm Sailors—*Oceanodroma*.

Eighth Group—Storm Divers, or Puffins—(*Puffini*)—One Family.

1. Diving Storm Birds—*Puffinus*.

SIXTEENTH ORDER.

RUDDER-FEET—(*Steganopodes*)—Four Groups.

First Group—Fishing-Plungers—(*Piscatrices*)—Two Families.

1. Tropic Birds—*Phaëton*.
2. Gannets—*Sula*.

Second Group—Frigate Birds—(*Tachypetes*)—One Family.

1. Frigate Birds—*Tachypetes*.

Third Group—Cormorants—(*Halieus*)—Two Families.

1. Anhingas, or Snake-neck Birds—*Plotus*.
2. Cormorants—*Phalacrocorax*.

Fourth Group—Pelicans—(*Pelecani*)—One Family.

1. Pelicans—*Pelecanus*.

SEVENTEENTH ORDER.

DIVERS—(*Urinatores*)—Six Groups.

First Group—Grebes—(*Podicipites*)—Two Families.

1. Grebes—*Podiceps*.
2. Dwarf Grebe—*P. minor*.

Second Group—Divers—(*Colymbi*)—One Family.

1. Sea Divers—*Colymbus*.

Third Group—Guillemots—(*Uriæ*)—Three Families.

1. Sea Doves—*Cepphus*.
2. Guillemots—*Uria*.
3. Crab Divers—*Mergulus*.

Fourth Group—Ornament Divers—(*Phaleres*)—One Family.

1. Ostrich Divers—*Phaleris*.

Fifth Group—Auks—(*Alcæ*)—Three Families.

1. Puffins—*Mormon*.
2. Auks—*Alca*.
3. Stump Auks—*Pinguinus, Plautus*.

Sixth Group—Fin Divers—(*Apterodytæ*)—Three Families.

1. Fin Divers—*Apterodytes*.
2. Fin Divers—*Sphenicus*.
3. Springing Divers—*Eudeptes*.

EXPLANATION OF PLATE II. (*Terminology of a Bird.*)

a. Upper mandible. b. Lower mandible. c. Nostril. d. Wing of upper mandible. e. Culmen, edge of lower mandible. f. Angle of the mouth. g. The eye. h. Front. i. Crown. k. Occiput, or hind head. l. Neck. m. Sides, or auditory part. n. Chin. o. Throat. p. Breast, or fore part of belly. q. Middle part of belly. r. Under tail coverts. s. Vent part of back. t. Middle part of back. u. Hind part of back. v. Tail feathers. w. Middle tail feathers. x. x. Side feathers of tail. y. Upper tail coverts. z. Under wing. aa. Shoulder wing coverts. bb. Wing coverts. cc. Primaries. dd. Secondaries. ee. Tertiaries. ff. Shoulder feathers. gg. Shank. hh. Tarsus. ii. Claw, or hind toe. kk. Outer toe. ll. Middle toe. mm. Inner toe. nn. Sole of foot.

FIRST CLASS.

CRACKERS—(*Enucleatores*)—THREE ORDERS.

Parrots being the most proportionally developed birds, ought to be placed at the head of the scientific classification. The question then is, what kind of birds should stand next? Reichenbach regards, as the next akin to Parrots, Sparrows proper (*Passeres*) and Ravens (*Coracirostres*). Not without reason, we call a certain Sparrow a "Pine Parrot," and certain Parrots (*Sittaces*) "Sparrow Parrots." This is founded on a similarity recognised as existing between those birds.

We must not, however, forget that, with such a conception, we have to do, in regard to such relations, only with orders, but not with groups or families; and that it is consequently the form only which we have to elucidate. A Cockatoo, a Finch, and a Crow appear to us as having no direct relation to each other; while all the Parrots, the multitude of Sparrows, and the Ravens show a decided similarity, each to the other. These three groups have many peculiarities in common. They are all short in body, with wings of a medium length; they have short, stout legs, a proportionally large head, and a short, hook-shaped, or a simple conical beak. Their tail, as with birds in general, is variously formed; it may be long or very short, gradually pointed or lyre-shaped; but it always consists of proportionally soft and somewhat elastic feathers; sometimes a peculiar formation of single feathers of the tail is found, but others a rich development of the tail coverts. The other plumage is compact, although not abundant; the single plumes are usually large and stiff; the color of the plumage is often vivid and frequently gorgeous. The anatomical structure is essentially similar in all. The skeleton may be said to be coarsely built, and the muscles are powerful. The tongue is of medium length, and capable of being little protruded or not at all, but very movable. The throat widens in many of the class and forms a regular crop; the stomach consists of strong muscles; and the senses are uniformly well developed. Sight, hearing, and feeling are developed in all; smell is developed in some, and taste to a certain degree in others.

Boldly, as well as in regard to instinct or intellect, we must rank the Crackers among highly gifted birds. They are wise, sprightly and quick in motion, and are fond of the society of their own kind. Their highly developed instincts or intellects enable them to live a comparatively comfortable life, even under unfavorable circumstances; while their bodily equipments afford them decided advantages in the struggle for existence. Parrots are chiefly confined to the warmer zones; the rest of the class are dispersed over the entire globe; their fixed quarters are essentially conditioned by the growth of trees, as by far the greater part of Crackers are tree birds. They rove about in comparatively small districts, and only those that live in colder regions are wanderers. Crackers live principally on vegetable food; their strong bill enables them to crack hard seeds which can not be eaten by others; they eat, besides, fruits and buds of trees or shrubs. Insects serve as food for many of them.

Nearly all Crackers mate for life, and most of them breed more than once a year. Their nests are variously constructed, and the number of eggs is usually from three to six or eight. The female generally does the hatching alone, but in some kinds she is temporarily relieved by the male bird. The feeding and raising of the young are equally conducted by both parents. Many Crackers are disliked by the farmer on account of their raids upon his property, but they usually profit him more than they do him damage. By their picking out the eyes of the seeds of weeds, and by their catching obnoxious insects, they become very useful to the husbandman. Besides, they enliven the woods and fields by their presence, their beauty, and their song. They can be tamed and kept in confinement with ease. The flesh of most of the Crackers furnishes healthy and delicious food, while the feathers of many of them are used as ornaments.

FIRST ORDER.

The Parrots—(*Psittacini*)—Six Groups.

The anatomical construction of Parrots has many peculiarities. Among these are the joint between the upper mandible and the front, the fully closed borders of the eye-sockets, the large bones of the palate, and the small neck-bone, which is, in some kinds, as in the Sparrow Parrots, entirely wanting. The breast-plate is quite large, and its high comb projects only a little. Among the soft parts the most remarkable organ, perhaps, is the thick fleshy tongue. The throat widens gradually to a crop, and the glandulous fore-stomach is separated by a smooth space from the stomach proper. The gall-bladder is wanting; the intestines are usually double the length of the body; the milt is small, and the kidneys thin-lobed. The wind-pipe, on the lower larynx, has three pairs of muscles.

The bill of a parrot bears some resemblance to that of a Hawk; but it is considerably thicker, stronger, and better formed. The nostrils are in the upper part of a cere or wax-skin, and are perfectly round.

The legs are short, strong, and fleshy; the toes are shorter than the middle toe, and covered with scaled only; the proportionately long toes have a thick sole; two toes are directed forward, and two backward; the claws are not long but gently bent, and are never very strong, yet tolerably sharp-pointed. The bones of the wings are of medium length and very strong; the large feathers are unusually numerous, but seldom long, and yet so arranged that the spread wing appears very pointed. The plumage consists of but few feathers, that, with the exception of the head feathers, are distinguished by their compactness. The feeble round the eye is naked; but the space between the eyes and the bill is generally feathered. A more or less brilliant leaf green is the predominant color, although there are hyacinth blue, purple, golden, yellow, and more darkly colored Parrots.

All the five senses are uniformly well developed in the Parrots. The Falcons are distinguished by keenness of sight, Owls by that of hearing; Ravens by that of smell, and Ducks by their discrimination of taste.

Parrots are considered by some as holding a rank among birds similar to that of apes and monkeys among quadrupeds. Like the latter they are imitative, cunning, and mischievous. All the larger kinds of Parrots live on the wing with considerable difficulty, but their flight is speedy. The smaller ones, for example, the Grass or Sparrow Parrots, fly with the swiftness of Swallows; the Araras and Paroquets fly quickly, and only the true Parrot fly rather slowly, flapping their short wings in order to propel their plump and heavy bodies. On the ground, Parrots move stiffly, and their walk is a mere waddle, although some of the ground Parrots run like Sandpipers.

They fly over long distances, and climb over narrow spaces. In the latter case they seat themselves with their bills, while other climbing birds use their feet only. Parrots seem no worse than a Raven, and can not dive at all. Their feet are like birds, and their bills, which, in most birds, serve instead of hands, are in Parrots more flexible than in other members of their class. From the use of the bill for the purpose of climbing, which a peculiar to Parrots, the kind is there is called the crossbill or Pine Parrot. The voice of the Parrot, though generally harsh and unpleasant, is often flexible and expressive. Several of the smaller ones of the Parrots sing to their female mates in such a charming way as entitles them to be ranked among warblers. Often kinds may be taught to whistle an air, and their capacity for imitating the human voice is the pronunciation of words is well known.

Parrots inhabit, exclusive of Europe, all the great divisions of the globe, but are principally found in the warmer regions. All

American kind is found as far north as the 42d degree of north latitude; another kind in the dreary deserts of Terra del Fuego, in the 53d degree of south latitude. Cockatoos harbor in New Zealand, and in the Macquarie island, under the 52d degree of southern latitude. In China, they are found only below the 27th degree northern latitude, and in India only up to the foot of the Himalaya; in Eastern Africa, they seldom pass northward beyond the 15th, or in Western Africa beyond the 16th degree of northern latitude. They generally, but by no means exclusively, confine themselves to the woods, as some kinds inhabit the treeless plains. Others live on the Andes above the wood region, and at a height of 11,000 feet above the level of the sea. Parrots, except in the breeding season, live in society, often in very large flocks; their regular settlements are made in forests, and their only roosts are large districts. They leave their sleeping-places early in the morning, and alight on the same tree to eat at no time; they set out side by side, and at their first warning cry take to wing. They arrive together at the same sleeping-place, and use it in common. The place is sometimes a hollow tree, a closely leaved tree top or a hollow in a rock. They seem to choose the closely leaved tree tops also for hiding-places. During a heavy tropical tempest, Parrots may be seen anxiously sitting on the highest dry branches of a tree, talking cheerfully to each other, while the storm is rushing down their bodies. As soon as it rains is over, they immediately dry and clean their plumage. The color of their plumage is so much like that of the foliage of the tree on which they are hidden, that it is difficult to see a single Parrot, although there may be fifty of them concealed among the leaves. If one of their number perceives an enemy approaching, he gives a loud, subdued warning cry, and the loud chattering of the whole company is at once silenced. Then they try to quit by windows climbing the side opposite the enemy; fly just as noiselessly away, and only begin to scream when about a hundred yards off; unseen; it would seem, to mock the deceived enemy than for any other purpose. Such a blind-man's-buff they regularly play, when engaged in robbing a tree of its fruit. All their thievish depredations seem to be generally executed with a similar cunning and mockery.

The food of Parrots consists chiefly of fruits and seeds. Many of the Loris feed exclusively on the honey of flowers. The Araras and Paroquets live also on fruits and seeds, and on the buds and flowers of trees and shrubs. Some Cockatoos add to these the larvæ of insects and worms. After feeding they fly to the water to drink and bathe; they drink a great deal, and sometimes even salt or brackish water; sometimes they are seen to bathe in the dry sand like chickens. They seem to be very fond of salt, for they are always found about the salt-licks in the forest. The time of incubation is in the season corresponding with our spring. The larger kinds breed but once a year, and never lay more than two eggs. The Australian Grass Parrots, however, lay from three to six eggs, sometimes even eight or nine, and breed twice, and often three times, in one year. The Sittiches and Cockatoos lay from three to four eggs, and breed once a year. The eggs are always white, with a smooth shell, and nearly round. Their nests are chiefly in hollow trees, but some of the American Parrots breed also in the hollows of rocks.

The Sittiches of India, according to Jerdon, often build their nests in the hollows in the walls of old buildings. The ground Parrots lay their eggs on the bare ground. All Parrots prefer to build their nests in large societies, sometimes in great multitudes. Sometimes a Parrot finds a hollow in a tree, but the entrance into it is too narrow, perhaps made by a small Woodpecker. The female widens the hole with her bill, so that she can inspect the inside; if this suits her, she widens the hole still more, hanging like a Woodpecker on the bark, and gnawing rather than cutting with her bill, till the hole is completed, which is sometimes the work of several days. The main thing in the construction of the nest is the hollow; a few chips on the bottom form a sufficient